SCIENCE AND BUSINESS OF CARBON FORESTRY

SCIENCE AND BUSINESS OF CARBON FORESTRY

H S Gupta • M Yadav • M Verma
A David • U K Sarma • C P Kala

teri
The Energy and Resources Institute

ISBN 978-81-7993-462-3

Suggested citation
Gupta, H. S., M. Yadav, M. Verma, A. David, U. K. Sarma, and C. P. Kala. 2014. *Science and Business of Carbon Forestry.* New Delhi: TERI

Published by
The Energy and Resources Institute (TERI)
TERI Press
Darbari Seth Block
IHC Complex, Lodhi Road
New Delhi – 110 003
India

Tel. 2468 2100 or 4150 4900
Fax 2468 2144 or 2468 2145
India +91 • Delhi (0) 11
Email teripress@teri.res.in
Website www.teriin.org

Printed in India

FOREWORD

There is now almost a universal consensus among climate change scientists that the world is warming up; the majority of them also believe that this warming is largely due to human-induced atmospheric emissions of greenhouse gases such as carbon dioxide. To help deal with the problem, the international community has resolved through the Kyoto Protocol of the United Nations Framework Convention on Climate Change, under which emission-reduction targets are set for developed countries and mechanisms are devised to help countries meet these targets. Moreover, there are other regional regulatory and voluntary efforts of different scale and level as well to assist the fight against climate change through a similar approach.

The Clean Development Mechanism (CDM) is one of the mechanisms of the Kyoto Protocol, which came into effect in February 2005. The CDM is an arrangement which allows Annex I countries (developed countries of the world) to invest in emission-reducing projects in developing countries as an alternative to what might be more costly emission reductions in their own countries. However, given weaknesses in the institutional and legal arrangements for afforestation and reforestation carbon projects, it is not surprising that progress in it has so far been minimal in many tropical countries, including India. The CDM project cycle for afforestation and reforestation projects is also comparatively demanding on project developers, involving cumbersome and time-taking project design and development, validation, registration, monitoring, verification and certification, and the issuance of carbon credits. It also poses challenges in the development and use of methodologies to define baselines, monitoring, and additionality. However, afforestation and reforestation carbon projects offer various valuable ecosystem products and services such as habitat for millions of species, timber, non-timber forest products, and supply of fresh water are also crucial in the global carbon cycle. Similar kinds of benefits are offered by the forestry carbon projects in voluntary mechanism also. At the same time, all the forestry carbon projects are critically important in regulating the global climate, besides offering various co-benefits unlike other type/sectoral projects.

To promote primarily forest restoration and other ways of forest management for carbon sequestration and carbon trade, a clear understanding of mechanism process and other aspects of the afforestation and reforestation CDM projects and other voluntary mechanism forestry projects is a must. Consequently, potential forestry carbon project developers in tropical countries require, among other things, full access to relevant information, tools, technologies, and knowledge of finance, economics, and markets for the appropriate design and outcome of project activities. To achieve this, the present book attempts to cater primarily to the needs of project proponents who want to build capacity with an aim to develop and implement such forestry carbon projects. In addition, this book also attempts to provide holistic guidance to foresters, environmentalists, tree growers, forest-based industries, academics, and students by providing related knowledge and information about climate change and forestry carbon projects, with respect to marketing, finance, economics, and other related aspects. The uniqueness of this book lies in being practice-oriented in nature. It contains a range of chapters dealing with the social, economic, and legal political aspects of forestry and climate change, exemplifying the diversity of approaches to forestry sector in a way never seen before in the Indian context.

I congratulate the editorial team and contributing authors for their excellent work in doing so. I specially appreciate Dr H S Gupta and his colleagues at the Indian Institute of Forest Management Bhopal for their contribution. We hope that this important book will serve as a useful tool to professionals serving in government, private, and non-governmental organizations, helping them to fully realize the potential of forestry carbon projects and thus will assist in preparing and developing forestry carbon projects, which will consequently deliver resources by way of rehabilitation of degraded forest along with sustainable development in general and particularly in Indian context. The present work also assumes great significance as it is fructifying in the "International Year of Forestry".

Dr P J Dilipkumar
Director General (Forest)
Ministry of Environment and Forests
Government of India

PREFACE

Carbon trading related to Clean Development Mechanism (CDM) is a cooperative mechanism conceptualized and established under the Kyoto Protocol. Along with some voluntary mechanisms, it has the potential to assist developing countries in achieving sustainable development by promoting environment friendly investment from industrialized country governments and businesses. In particular, CDM is a project-based mechanism that allows public or private entities from countries with emission reduction targets (Annex I countries) to invest in greenhouse gas (GHG) emission reduction projects in developing countries, to earn carbon credits, known as certified emission reduction (CER) units. These credits can be used against domestic emission reduction targets assigned to Annex I countries or can be sold to other interested parties on a voluntary basis.

Of the various kinds of projects allowed under the regulatory and voluntary mechanisms, afforestation and reforestation projects are of interest. A great deal of work has been done to identify opportunities for afforestation and reforestation carbon sink, particularly in the rehabilitation of degraded tropical forest lands. India can host numerous such projects as it is a Non-Annex I country and thus has ample opportunities to reduce emissions.

A private sector consultancy firm active in carbon markets estimates somewhat more conservatively that the worldwide supply of CERs could average between 150 and 250 million tonnes per year over the first Kyoto commitment period (2008–12), which could cover up to half of the predicted global emission reduction shortfall against Kyoto targets (Sharma 2010).

According to the Intergovernmental Panel on Climate Change report on Land-use, Land-use Change and Forestry (LULUCF), there is the potential for the forestry sector to absorb 12%–15% of global emissions originating from fossil fuels.

It is expected that the LULUCF sector in general and forest departments, agriculture departments, or many related natural resource departments in particular in India should be in a position to assess the potential to use the degraded and low-productivity lands to promote forestry-related activities through raising plantations, afforestation, and reforestation. Such an initiative, if taken by government departments, private companies, and individuals and communities, could help them accrue large amount of carbon credits and may also contribute to India's vision of achieving 33% forest cover, as well as providing sustainable livelihood.

This book is meant for persons and institutions who work for forestry and tree-farming activities and foresters (both government and non-government) with an interest in forest carbon projects. It aims to facilitate setting up forest carbon projects and let the whole society benefit. This book is structured covering major aspects: including the background of climate change and forestry, introduction to the carbon market; forest carbon project development and the timeline; economics of the projects; sources of different funds and grants and other financial aspects are the major topics.

This book is expected to be very useful to the reader given the very fast changing economic-political framework resulting in changes in legislations, as well as actors, funds and regulations. The lucid explanations and coverage of different topics associated with forest carbon projects would certainly help all the readers in general.

This book arises from the long professional and practical experience gained by an interdisciplinary team of technical experts working in the Indian Institute of Forest Management Bhopal, having both academic and field experience of national and international level.

The present book is expected to fill up the knowledge gap for practitioners and anybody associated with designing, formulating, and finally operating such projects on forestry carbon.

H S Gupta

ACKNOWLEDGEMENTS

The authors sincerely thank the following for their support and encouragement in writing this book in the form of their inputs and critical comments for improvement of the book: Dr P J Dilipkumar, Director General of Forest, Ministry of Environment and Forests (MoEF) (Government of India); Shri J Kishwan and Shri A K Bansal, both Additional Director General of Forest, MoEF; Shri K B C Thampi, former Inspector General, National Afforestation and Eco-development Board (NAEB); Shri Ashok Pai, former Deputy Inspector General, NAEB; Shri D K Sharma, Deputy Inspector General, NAEB; and Shri A K Singh, Principal Chief Conservator of Forests, Jharkhand Forest Department. The authors are thankful to the sincere contributions made by P Upadhyay (PGDFM), who has worked hard through his inputs as well as editing.

This book would not have been possible without the financial support provided by the NAEB, and the editors and authors sincerely thank the NAEB for this support and having reposed faith in the academic acumen of the Indian Institute of Forest Management (IIFM) Bhopal.

We also gratefully acknowledge the many academicians and practicing foresters who have helped us in structuring the ideas with their suggestions and data that laid the foundation for this book. The authors also thank numerous friends and well-wishers for their formal and informal comments on this draft and related documents. They deserve special thanks for this support in the development of this analysis by their incisive criticisms at the draft level.

Although all the attempts have been made to acknowledge these experts in the text and in the reference, no words are adequate to express the gratitude to them. The authors sincerely express their gratitude for reproducing the work of these experts and thus bringing their work before the readers of developing countries such as India. Hence, no originality of thoughts and concepts is claimed in this book and all the thoughts, concepts, and research cited by the distinguished experts are again duly acknowledged.

The authors also thank Dr R B Lal, Director of IIFM Bhopal, for reviewing the work and offering helpful comments and suggestions, and above all for his constant support and guidance on the overall work.

Thanks are also due to the various state forest departments, particularly of Jharkhand and Madhya Pradesh, and their officers who have always supported and encouraged this work.

H S Gupta
M Yadav
M Verma
A David
U K Sarma
C P Kala

DISCLAIMER

This work does not represent or vouch for the accuracy and suitability of the content of the chapters or the conclusions as set out therein. It is the sole responsibility and obligation of the reader of this book to satisfy himself/herself about the accuracy and suitability of the information. The authors shall have no liability for any inaccuracy, representation, or misrepresentation set out herein. The reader further agrees to hold the editors and authors harmless from and against any claims, loss, or damage in connection with or arising out of any commercial decisions made on the basis of the information contained herein. The reader of this book is strongly advised not to use the content of this report in isolation, but to take the information contained herein together with other information and to formulate his/her own views, interpretations, and opinions thereon. The reader is strongly advised to seek appropriate legal and professional advice before entering into business transactions.

The objective of this work is to facilitate the wide range of stakeholders, meeting their information needs, and improving their conceptual understanding in a broad framework, which has got national and societal relevance. The authors do not have any commercial interest in the work and claim no originality to any idea, concepts, and information and acknowledge all the sources, quoted and adapted in this work, which is expected to be socially useful.

Contents

Appendices

Acronyms and Abbreviations

AAU	assigned amount unit
AB32	California's Global Warming Solutions Act
ACTIVE	Action for Collective Tribal Improvement and Vocational Education
ADR	alternative dispute resolution
AFOLU	agriculture, forestry, and other land use
AGRA	Alliance for a Green Revolution in Africa
ARB	Air Resources Board
ASB	alternatives to slash and burn
BAU	business as usual
BEE	Bureau of Energy Efficiency
BioCF	BioCarbon Fund
CAR	Climate Action Reserve
CBFF	Congo Basin Forest Fund
CCAR	California Climate Action Registry
CCB	climate, community, and biodiversity
CCX	Chicago Climate Exchange
CDCF	Community Development Carbon Fund
CDM	Clean Development Mechanism
CER	certified emission reduction
CGIAR	Consultative Group on International Agricultural Research
CIAT	Centro International de Agriculture Tropical (International Centre for Tropical Agriculture)
CIF	Climate Investment Funds
CIFOR	Centre for International Forestry Research
CME	coordinating and managing entity

CMP	Conference of the Parties Serving as the Meeting of the Parties to the Kyoto Protocol
CO_2e	carbon dioxide equivalent
COP	Conference of Parties
CORE	Carbon Offset Research and Education
CPA	component project activity
CPA-DD	component project activity design document
CPCB	Central Pollution Control Board
CPF	Collaborative Partnership on Forests
CSR	corporate social responsibility
DBH	diameter at breast height
DCF	Danish Carbon Fund
DNA	Designated National Authority
DOE	Designated Operational Entity
DOM	dead organic matter
EB	Executive Board
EMS	environmental management system
ERPA	emissions reduction purchase agreement
ERU	emissions reduction unit
ET	emissions trading
ETF-IW	Environmental Transformation Fund—International Window
ETS	Emissions Trading Scheme
EU	European Union
EUETS	European Union Emissions Trading System
FAO	Food and Agricultural Organization
FCPF	Forest Carbon Partnership Facility
FORMA	Forest Monitoring for Action
GDP	gross domestic product
GEF	Global Environment Facility
GGAS	Greenhouse Gas Abatement Scheme
GHG	greenhouse gas
GIS	geographic information system
GOI	Government of India

GPS	global positioning system
GTZ	Gesellschaft für Technische Zusammenarbeit
ha	hectare
HOPE	Homeless Outreach Poverty Eradication
ICF	Italian Carbon Fund
ICRAF	International Centre for Research in Agroforestry
IFCI	International Forest Carbon Initiative
IIFM	Indian Institute of Forest Management
IMD	India Meteorological Department
INIA	Instituto Nacional de Investigación Agraria
IPCC	Intergovernmental Panel on Climate Change
ISO	International Organization for Standardization
JI	Joint Implementation
lCER	long-term certified emission reduction
LCF	locally controlled forestry
LoA	letter of approval
LoI	letter of intent
LULUCF	land use, land-use change, and forestry
MCX	Multi Commodity Exchange
MDG	Millennium Development Goals
MEA	millennium ecosystem assessment
MoF	Ministry of Finance
NAEB	National Afforestation and Eco-development Board
NAMA	Nationally Appropriate Mitigation Action
NAPCC	National Action Plan on Climate change
NCDEX	National Commodities and Derivatives Exchange
NEDO	New Energy and Industrial Technology Development Organization
NETS	National Emissions Trading Scheme
NGACs	NSW Greenhouse Abatement Certificates
NGO	non-governmental organization
NMEEE	National Mission on Enhanced Energy Efficiency
NSW	New South Wales
NTFPs	non-timber forest products

NYMEX	New York Mercantile Exchange
OC	opportunity cost
OCT	Oregon Climate Trust
ODA	Official Development Assistance
OECD	Organization for Economic Cooperation and Development
OTC	over the counter
PAT	Perform, Achieve, and Trade
PDD	project design document
PES	payments for ecosystem services
PIN	project idea note
PoA	programme of activities
PoA-DD	programme of activities design document
PPCR	Pilot Programme for Climate Resilience
PSPD	Paperboards and Specialty Papers Division
QA/QC	quality control/quality assurance
REDD	Reducing Emissions from Deforestation and Forest Degradation
RGGI	Regional Greenhouse Gas Initiative
RIL	reduced impact logging
SCF	Strategic Climate Fund
SCS	Scientific Certification Systems
SERP	Society for Elimination of Rural Poverty
SFM	sustainable forest management
SI	Systéme International
SIDS	Small Island Developing States
SNIRD	Society for National Integration through Rural Development
SOC	soil organic carbon
SREP	Scaling-up Renewable Energy Programme for Low Income Countries
TAR	*Third Assessment Report*
TBD	to be determined
tCER	temporary certified emission reduction
tCO_2	tonnes of carbon dioxide

tCO_2e	tonnes of carbon dioxide equivalent
TIST	International Small Group and Tree Planting Programme
TNC	The Nature Conservancy
TSBF	Tropical Soil Biology and Fertility Institute
UNCED	United Nations Conference on Environment and Development
UNFCCC	United Nations Framework Convention on Climate Change
UNU	United Nations University
VCS	Voluntary Carbon Standard
VER	verified or voluntary emission reduction
VOS	voluntary offset standard
WBCSD	World Business Council for Sustainable Development
WRI	World Resource Institute

1

IMPLICATIONS OF CLIMATE CHANGE

H S Gupta

INTRODUCTION

Climate change has emerged as one of the most important global environmental challenges because of its wide implications on food production, freshwater supply, health, and natural ecosystems (Figure 1.1). Given its importance, the United Nations has included it in the Millennium Development Goals. The United Nations Framework Convention on Climate Change (UNFCCC) defines climate change as "a change of climate which is attributed directly or indirectly to human activity that alters the composition of the global atmosphere and which is in addition to natural climate variability observed over comparable time periods". Today, people also use it as a synonym of global warming.

Climate change can be defined as a long-term change in the state of the weather pattern, as validated by statistical tests and historical evidence, and measured by the changes in mean and variability of its properties that persist for an extended period. Climate change can be attributed to modifications in the atmosphere, such as changes in the composition of gases. The causal factors for these changes are of two types: anthropogenic (such as fossil fuel burning and deforestation) and natural (such as changes in solar radiation and volcanism).

There is enough scientific evidence that establishes the linkage between greenhouse gas (GHG)[1] emission and climate change. For

[1] Six greenhouse gases are mainly responsible for global warming: carbon dioxide, methane, nitrous oxide, chlorofluorocarbons, hydrofluorocarbons, and perfluorocarbons.

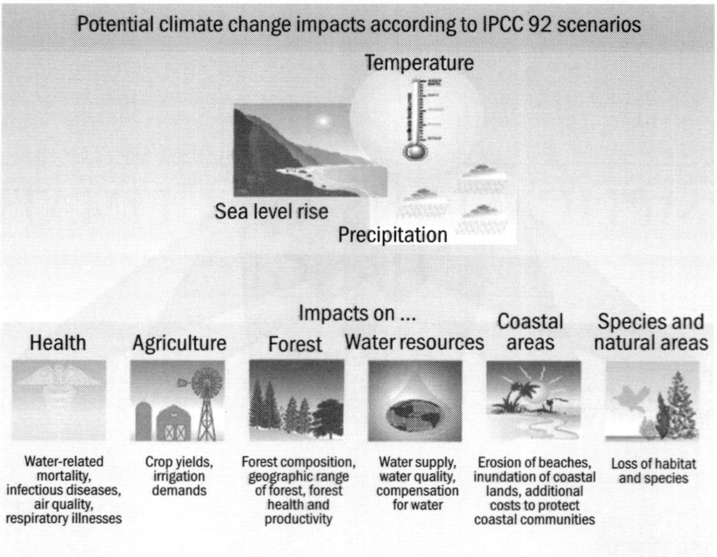

FIGURE 1.1 | Climate changes and its impacts

Source: IPCC (2001)

instance, it has been proved that the warm years experienced during the last 50 years resulted from extensive emissions induced by human beings in pursuit of economic and industrial development (IPCC 2001).

HISTORY OF CONCERN FOR CLIMATE CHANGE

Although awareness about climate change had been there for quite some time, it became prominent in 1974 when Frank Sherwood Rowland and Mario J. Molina published a paper hypothesizing that chlorofluorocarbons (CFCs) contribute to ozone depletion. The Montreal Protocol further emphasized the issue. The environmentalists started working out the relationship between carbon emissions and climate change. Consequently, 172 countries under the United Nations participated in the Earth Summit held in Rio de Janeiro and signed the UNFCCC. In 1997, they adopted the Kyoto Protocol to stabilize the overall anthropogenic GHG emissions in the atmosphere. By that time, it was established that unsustainable exploitation of natural resources for industrialization and development across the world is the main cause of climate change. As a result, reduction of GHG emissions was made mandatory under the Kyoto Protocol for the industrialized/developed nations, which contribute more to environmental pollution in comparison to developing countries. Some schools of thought believe that this problem is caused by the exploitation of the common-pool resources by developed countries, which release enormous emissions into the atmosphere without any restrain. This situation is aggravated by the decrease in the natural carbon sink, such as forest, as a result of expanding human population. All these factors led apex world bodies to impose some restrictions (currently through the Kyoto Protocol) on the callous attitude of the developed world, thereby alleviating the problems of environmental pollution and global warming.

EFFECT AND ECONOMIC CONSEQUENCES OF CLIMATE CHANGE AT GLOBAL LEVEL

Global warming induced by climate change has widespread economic consequences on human life, mainly in developing countries or tropical regions (Kelman and Jennifer 2009). The people living in developing countries either do not have resources to combat climate change or are not aware of its adverse ecological and economic impacts despite climate change being an inevitable truth that will affect their life physically and economically (Stern 2007). If appropriate strategies to combat climate change are implemented adequately, then the impacts of global warming can be reduced. However, there has been uncertainty about the severity of the impacts of climate change until now. In the context, many models have been developed to predict the impacts of climate change. But these models are not very accurate because of the distinction between the effects at the global and regional levels and their integration. Also there is not enough evidence to understand directly or indirectly the responses of individuals, firms, and governments towards climate change (IPCC 2007).

An Intergovernmental Panel on Climate Change (IPCC) report suggested a mean temperature rise of 1.4–5.8°C along with an estimated accumulation of GHGs in the atmosphere to the level well above 550 parts per million (ppm) CO_2 equivalents by 2100 (IPCC 2001). This unprecedented rise may have severe implications on hydrological systems, ecosystems, crop production, rise in sea level, and forest ecosystem as well, particularly in tropical areas, which mainly consist of developing countries, including India. The cost associated with the strategies planned to combat climate change has been estimated to be around 0.5%–1% or may be as high as 2% of the total global gross domestic product (GDP) per annum (Stern 2007), which is very high. In extreme scenarios, Small Island Developing States, such as Maldives, may have to fight for their existence (Sathaye, Shukla, and Ravindranath 2006). India, in addition to its domestic problems, may face stiff immigration problems if the sea level rises, which may also cause inundation of one-fifth of Bangladesh.

As stated earlier, the direct effects of climate change, such as increase in temperature, change in hydrological cycle, sea-level rise, and famines, will be more prominent in poorest countries where people are already suffering from various problems. For example, low-lying coastal plains, where the economy is heavily dependent on rice cultivation, have the highest flood risk due to sea-level rise. Intrusion of salty water due to sea-level rise will destroy the capacity of these lands and make them unsuitable for rice cultivation, leading to scarcity of food in the affected countries, including India (IPCC 2007). On the other hand, indirect effects will be more drastic in terms of affecting the ecological services that human beings draw from the environment. These effects include erratic rainfall, increase in the risk of vector-borne diseases due to rise in temperature, impact on marine ecosystems due to acidification of the sea, vulnerability of small rivers and tributaries, and loss of biodiversity. Similarly, increased shortage of drinking water will pose serious problems for two-thirds of the world population by 2025. The countries that already suffer from shortage of water will feel the pinch without the assistance from other countries (Sathaye, Shukla, and Ravindranath 2006).

Although no comprehensive model has so far been able to accurately predict the cost associated with climate change, many studies suggest that it will be huge for developing countries. By 2100, the effect of climate change may be reflected in the lowering of India's and South East Asia's GDP by 9%–13%. An estimated 145–220 million people may have to survive in less than $2 a day, which may increase malnutrition and child deaths (Stern 2007).

INDIA'S CONCERN

India's agricultural and forest systems depend mostly on monsoon-induced rainwater. Scientists predict that rain system will become more inconsistent and erratic, causing unprecedented damage to the yield from agriculture and forest. This places India in a very critical situation.

Forest ecosystems are critical for environmental sustainability; they offer quality of life to local communities and national economies, particularly in developing countries. The forestry sector plays an important role in controlling global CO_2 control, reducing emissions, and removing the CO_2 accumulated in the atmosphere over the years and restoring it in soil, vegetation, and wood products. Further, the forestry sector is closely linked to socio-economic systems, particularly for forest dwellers and rural communities in developing countries. The projected climate change due to increased GHG concentration in the atmosphere will affect forest ecosystems, biodiversity, regeneration, biomass growth rate, and geographic distribution of plant species (Ravindranath 2006).

The coastal areas will be more vulnerable because they are more prone to inundation and salinization. This negative change in agricultural and forest yield may result into loss of jobs from already over-employed Indian agricultural sector, along with reduced incentives for farmers to grow crops and practise agroforestry. Hydrological cycles may get disrupted and altered due to the increase in the severity of rainfall. In tropical regions such as India, a 20% rise in summer monsoon has been predicted, which will form the basis for frequent epidemics. In addition, there may be a shortage in the available run-off on the surface of earth, which will reduce the availability of drinking water (MoEF 2004). According to the IPCC, the sea level is expected to rise by around 38 cm until 2050. In the Indian scenario, the impact on low-lying areas such as Mumbai may cost more than ₹2000 billion towards inundation and rehabilitation (Kumar, Parag, and Shalini 2008). Health hazards may increase because of the rise in temperature and the relation between epidemics and infrequent weather conditions. Mud puddles formed by floods and by rivers in arid times become the breeding grounds for mosquitoes. In Gujarat and Rajasthan, excessive rainfall is often followed by malaria/epidemics.

India is home to many forest ecosystems containing a variety of exclusive flora and fauna, which are great sources of products and services for the Indian population. These products are in danger as a result of the change in the ecosystem induced by global warming. The loss of this biodiversity will pose extreme threat to forest ecosystems and their ecological values, and the opportunity cost connected with the loss will further result into poverty. For instance, the habitat of the already endangered Lesser Florican and Indian wild ass, found in Rann of Kutch (Gujarat), may get further reduced or lost due to rise in the sea level.

Thus, India, with a population of over a billion, is highly vulnerable to the climate change impacts. Therefore, to ensure the well-being of its people, India has to adopt some strategies towards the fight against the effects of climate change.

Although the developing world is indebted to the developed world, if the cost associated with climate change is taken into consideration, the industrialized nations will owe more than $600 billion to the developing nations for disturbing the balance of nature and forcing them to invest heavily to mitigate the adverse effects of climate change. This amount is three times the conventional debt that the developing countries owe to the developed ones. However, this debt should be considered an obligation of the developed countries towards the developing countries. The developed countries should now pay for their uncurbed exploitation of natural resources so far and for polluting the atmosphere, a universal resource in the past (Christian 1999). The West can contribute to the East and South through less energy-intensive technologies and by facilitating the adoption of the Clean Development Mechanism (CDM) with equity considerations. At the same time, the developing countries need to develop climate-friendly technologies for sustainable growth. Equity issues, however, need to be addressed with respect to the transfer of energy-efficient technology from the developed countries to the developing countries.

In conclusion, one can say that climate change is an issue that should be used to establish economic equity across the nations and equity in terms of utilizing universal resources. Unfortunately, these issues have been largely ignored in global discussions where rich nations usually dominate. The talks generally revolve around economic effectiveness and the participation of developing countries in emission cuts. Until now, the developed world, through CDM-like mechanisms, have used the land of the developing and poor nations to cut down their emissions, while exploiting universal resources. Hence, very few easy options are left for the developing countries.

SOLUTIONS TO THE CLIMATE CHANGE PROBLEM

As of now, climate change seems to be a reality with no immediate or long-lasting solutions. Even if all the present anthropogenic emissions were to stop, changes would still continue into the future. Therefore, there is an urgent need for increased action on climate change, through both mitigation and adaptation efforts.

Mitigation efforts can be defined as efforts that try to restrict the adverse impacts of future climate change by controlling its causes. They can also be described as human intervention to reduce GHG sources or enhance carbon sequestration. On the other hand, adaptation is a process of reducing the overall impacts of climate change on the environment. Adaptation to climate change also refers to adjustments in natural or human systems in response to climatic stimuli or their effects, which moderate harm or exploit beneficial opportunities.

Currently, enough scientific evidence exists that confirms the occurrence of climate change, but there is no consensus on its aftermath. Under the scenario of climate-change-induced disasters, the future of the flora and fauna resources of forests looks very uncertain and bleak. Thus, climate change mitigation is urgently required to save forest resources. Yet, mitigation cannot be the only solution because even strict measures for climate change mitigation across the globe cannot stop climate change.[2] Thus, an effective synergy between the mitigation and adaptation processes is needed to limit the undesired consequences of climate change. Although adaptation plans can also be helpful for affected people, mitigation campaign is a better option for long-term biodiversity conservation of forest and its resilience.

Although ecosystems have adapted to changing conditions in the past, current changes are occurring at faster rates that were not seen before. In general, the faster the climate changes, the greater is the impact on people and ecosystems. Reductions in GHG emissions can lessen these pressures, giving these systems more time to adapt. Biodiversity-based adaptive and mitigative strategies can enhance the resilience of ecosystems and reduce the risk of damage to human and natural ecosystems. The following activities can promote the mitigation of or adaptation to climate change (CBD 2007):

- Maintaining and restoring native ecosystems
- Protecting and enhancing ecosystem services
- Managing habitats for endangered species
- Creating refuges and buffer/protected zones

[2] Details available at <www.ittefaq.com>

- Establishing networks of terrestrial, freshwater, and marine protected areas that take into account the projected changes in climate

In the global climate change discussions, the issue of extreme importance to the developing countries is alleviating the vulnerability of their natural and socio-economic systems against the projected adversities of climate change. Although it is being keenly contested that both mitigation and adaptation are required to combat climate change, adaptation is a highly localized and specialized effort with limitations and boundaries. Moreover, adaptation does not promote sustainable development and environmental restoration, so for the time being mitigation efforts can prove really fruitful. Keeping this in mind, the Kyoto Protocol was designed (discussed in Chapter 2), which is one of the best mitigation efforts in the history of climate change. The Protocol is expected to deliver the following major benefits:

- Establishing equity between the developed and developing nations in the context of climate change
- Inclusive growth for local communities and sustainable development
- Ecological and environmental restoration

In a major response to the threat of climate change, the Kyoto Protocol will assist global regime to mitigate the widespread causes of climate change and also to set binding emission reduction targets. Thus, before the preset of the negative effects of climate change, mitigation efforts can assist in stabilizing the carbon content in the atmosphere, providing a better environment for all the creatures of the world to survive. However, besides the international regulatory efforts through the Kyoto Protocol, many efforts based on voluntary approaches are also being practised (involving both mitigation and adaptive strategies).

2

GLOBAL EFFORTS AND MECHANISMS TO COPE WITH CLIMATE CHANGE

H S Gupta

EFFORTS

Climate change is one of the most undesired consequences of unsustainable developmental practices being pursued all over the world, particularly since the advent of industrialization. Factors such as burning of fossil fuels, deforestation, land-use change, and modern agricultural practices have contributed to greenhouse gas (GHG) emissions. Solar radiations trapped by these emissions have led to global warming, which will have discernible impact on the global climatic system, due to rise in sea level and changes in weather patterns, and will have adverse health effects (Kelman and Jennifer 2009).

As discussed in Chapter 1, two broad strategies have been developed to curb the ill effects of climate change: mitigation and adaptation. Although adaptation reduces the impacts of climate change, it does not have much control over the causes of climate change. That is why the international regime concentrated on mitigation efforts in the beginning. It evolved a formal and obligatory mitigation strategy known as the Kyoto Protocol, along with many other voluntary commitments. With 192 countries endorsing the protocol, it is the most widespread, globally discussed, and accepted regulatory mechanism. On the other hand, voluntary mechanisms are less regulated and less stringent in terms of processes and procedures, and they depend on self-discipline. Being less

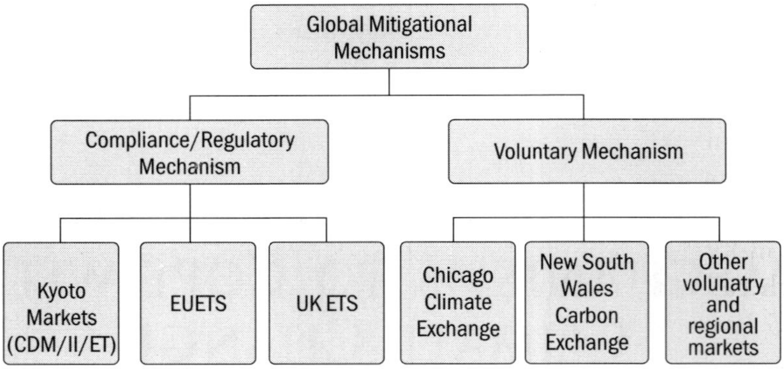

FIGURE 2.1 Global mitigation mechanisms

complicated, these mechanisms certainly motivate organizations and individuals to adopt cleaner technologies or to go carbon neutral.

There are two broad categories of mechanisms: regulatory and non-regulatory (Figure 2.1). This categorization is based on the global control of the concerned mechanisms. The United Nations Framework Convention on Climate Change (UNFCCC) controls the broadest regulated mechanisms, including the Kyoto Protocol, which is the global regime. Rules are formulated with the consent of all the member parties and nations. On the other hand, voluntary mechanisms are regionally regulated or self-induced and are followed by the organizations that are willingly contributing to the fight against global warming or climate change. The rules in the voluntary market vary according to the market and are highly specific.

KYOTO PROTOCOL

The United Nations Conference on Environment and Development in Rio de Janeiro led to the formation of the Framework Convention on Climate Change (FCCC), which eventually became the UNFCCC in a major response to tackle the threat of climate change at the global level. The convention came into force in 1994 with an objective of stabilizing the GHG level in the atmosphere. Initially, a legally non-binding, voluntary pledge was included: "The major industrialized/developed nations would reduce their GHG emissions to 1990 levels by the year 2000 and all nations would undertake voluntary actions to measure, report, and limit GHG emissions" (Susan 2001). However, it became apparent within the initial years that the voluntary stabilization target set by the UNFCCC would not be met by 2000 because major nations such as the USA and Japan would significantly fall short of this target (Susan 2001).

Consequently, the Conference of the Parties (CoP)[1] to the treaty decided in 1995 to enter into a legally binding (not voluntary) agreement, and negotiations began for a protocol to establish legally binding limitations or reductions in GHG emissions. During the CoP-3 meeting held in Kyoto (Japan) in 1997, the Kyoto Protocol was formally adopted, which calls for significant obligatory emission reduction to stabilize the GHG levels in the atmosphere.

Since developed countries have contributed to the major share (83% and 63% in the nineteenth and twentieth centuries, respectively) of GHG emissions through the combustion of fossil fuels (because of their long, unsustainable developmental process), they hold the prime responsibility for undoing climate change (Sathaye, Shukla, and Ravindranath 2006). Thus, it was decided in the CoP-3 that limitations would be set only for the developed countries, which were referred to as "Annex I countries"[2] (developing countries are referred to as "non-Annex I countries"). This was referred to as the "Berlin Mandate", which is a continuation of the principle established in the UNFCCC that parties bore "common but differentiated responsibilities and respective capabilities, social and economic conditions". Under this, industrialized (Annex I) countries were held responsible for the past GHG emissions and were set a target to reduce the emissions by 5.2% of the 1990 level until 2012. However, it took eight years for the Kyoto Protocol to come into effect in 2005, when Russia ratified it,[3] meeting the criteria of ratification by at least 55% of GHG-emitting countries.

The prime objective of the Kyoto Protocol (Figure 2.2) is to regulate global GHG emissions. To assist Annex I countries in attaining their target of emission reduction, the Kyoto Protocol has three flexible mechanisms: emissions trading, joint implementation, and Clean Development Mechanism (CDM).

These mechanisms are to be used as a supplemental option for achieving reduction target in addition to the domestic actions. Moreover,

---

[1] Conference of the Parties serving as the highest body in UNFCCC is constituted by the participation of different countries.

[2] The 38 developed nations, excluding the countries whose economies are in transition, are listed as Annex I in the UNFCCC mandate.

[3] Ratification is the process of accepting the protocol as a binding measure by the parliament of the respective country.

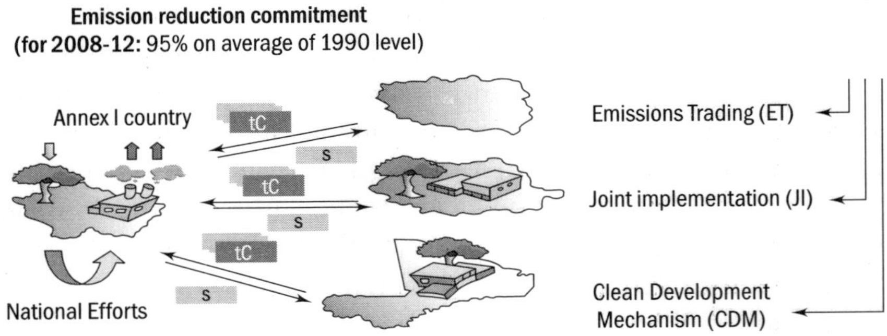

FIGURE 2.2 | Kyoto Protocol at a glance

these mechanisms are designed to promote cooperation and channelize funds between different nations, particularly between developed and developing nations. These are discussed in detail in the following subsections.

Emissions Trading

As set out in Articles 17 and 22 of the Kyoto Protocol, emissions trading provides opportunity for Annex I parties to acquire certified emission reductions (CERs) from other Annex I parties to fulfil their commitments. Any such action for a party shall be supplemental to the domestic action for meeting quantified emission reductions. This is executed by setting up an emissions limit for each party, known as assigned amount units, and permitting the party to trade a part of it. Apparently, a party can trade the emissions if it reduces the emission more than what has been legally asked or, in other words, if it saves a part of the emissions that it was assigned. For example, if a party A has been assigned an emission target of 100 units of GHGs, but it emits only 80 units, the 20 units that was saved can be traded with either an intermediary or a party that needs it. However, after all trading is finished, the net result will remain the same because the saved emission is being traded to a country that overshoots its target emission. In other words, trading redistributes the allowed emissions between different parties, while keeping the total emissions within the originally agreed limit.

This incentive mechanism of emissions trading has been intended to promote worldwide energy-saving measures and reduce energy wastage, consequently reducing GHG emissions in the process to produce extra energy. There are huge opportunities in countries where domestic and per capita consumption is high because this mechanism may offer much higher price than the actual cost of saving energy or reducing emissions. In the absence of such a mechanism, countries could have tried only to meet their obligatory emission reduction target instead of putting in some extra effort. Thus, this incentivizing mechanism provides enough motivation for organizations to cut down their extra and redundant emissions in pursuit of extra economic benefits. "Many economists believe this system to be an economically effective means of achieving global emission reductions at least cost. If the trading system works well and no one sells too many permits, then the total emissions from the Annex B Parties should be kept within the limit set originally for all Parties together. Such trading regime could therefore, in theory, bring about a more cost-effective reduction in global missions than would otherwise be the case and produce some other overall economic benefits. This would be the ideal situation, but there are many ways in which a global emissions trading regime may also go wrong" (Robertson 1998).

Current updates about the efficacy of this mechanism can be supported by the fact that "estimates by the Massachusetts Institute of Technology (MIT) that the European Union Emissions Trading Scheme has cut European emissions by 120–300 million tonnes of CO_2 ($MtCO_2$) during its first, highly imperfect phase—up to 5% of emissions from the covered sectors, despite excessive allocations of emissions allowances. It captured private sector attention like no other climate initiative by its rapid introduction and impact" (Grubb, Brewer, Misato, *et al.* 2009).

Joint Implementation

As defined in Article 6 of the Kyoto Protocol, under joint implementation, Annex I parties may implement a project that reduces emissions or removes carbon from the atmosphere in the territory of other Annex I parties, and in return earn emission reduction units. This joint project, as defined in the Kyoto Protocol, should be additional to what would otherwise have occurred. This mechanism was initially proposed in 1991 by the developed countries that voluntarily committed emission reduction through various innovative technologies and energy-saving measures. These countries believed that it was better to start a project with inbuilt capacity of energy-saving measures than to first build it and then implement energy-saving measures. The developed countries debated that it would be more cost effective for them to reduce emissions in other countries where new projects at very nascent stage provided vast opportunities for implementing energy-efficient technologies or introducing renewable energy systems, and also where reducing GHG emissions or curtailing its growth was cheaper than at home. Consequently, Article 6 of the Kyoto Protocol provides scope for developed countries, as categorized in Annex B of the Kyoto Protocol, to implement projects in other developed countries with an ultimate objective of reducing carbon emissions.

This mechanism can be executed through setting up of emission reduction projects in these countries, financed by developed countries. The "donor" country pays or finances the projects, while the project takes place in the "host" country. The emission reduction thus achieved helps the donor country to get the credit. As originally envisaged, joint implementation could be between two countries with targets or between a donor country with a target and a host country without a target (Robertson 1998). The very logical issue that lies at the core of this mechanism is that any new project will definitely add on to the current emissions. So the Kyoto Protocol focuses on the additionality aspect of the project, where a project can be called additional when it replaces another project that was environmentally unsustainable and was emitting larger emissions. This emission reduction achieved by replacing the project technology or by introducing renewable technologies will be called additional and can be accredited to the donor country's account.

Clean Development Mechanism

As defined in Article 12, CDM helps Annex I parties implement energy- and carbon-efficient projects in non-Annex I parties to reduce emissions or absorb carbon through afforestation and reforestation activities (which will earn CERs for the Annex I parties) and to assist the host parties in achieving sustainable development, which is the ultimate objective of the Convention. So far CDM is the only formal link between developed and developing countries in the context of climate change. CDM has great potential in terms of securing and channelizing funds from the developed countries into the developing countries, thus helping developing countries to access better and energy-efficient technologies to participate in the global fight against climate change. CDM also approves afforestation and reforestation projects (the only form of forestry projects included in the Kyoto Protocol), leading to the inclusion of forestry projects with their additional environment and social co-benefits.

Clean Development Mechanism is a scheme under an international climate change agreement that allows developed countries to buy "credits" from projects that supposedly reduce GHG emissions in the developing countries, instead of cutting their own emissions domestically. In doing so, the scheme claims to be mitigating climate change while also promoting sustainable development.

Current estimates show that CDM is the only mechanism that attempts to subsidize the developing countries for their environmental conservation measures by channelizing funds from developed countries to help them access low-cost carbon technologies and achieve sustainable development. CDM can help the developing countries avail approximately €17.5 billion (at the current market price of €10 per tonne of CO_2) to reduce around 1.75 billion tonnes of CO_2 during the period 2008–12 (Wara 2007). However, there are claims that CDM projects are not fulfilling their expected goals. Box 2.1 gives an example of deviation from the objectives in the Philippines.

OTHER PROVISIONS OF KYOTO PROTOCOL

The Kyoto Protocol contains provisions to enhance the capacity of mitigation measures with respect to countries and also promotes the countries to implement policies, subsidies, and measures to reduce carbon

BOX 2.1 ***Distortion at the ground: the concerns***

An evaluation of the existing Clean Development Mechanism (CDM) projects in the Philippines as of June 2009, however, raises questions as to whether the scheme is, in fact, undermining its own purported aims. It is apprehended that most of the "credits" being generated may go to projects that further exacerbate climate change and compromise sustainable development, enriching large conglomerates that are expanding extractive and fossil-fuel-intensive activities, in pursuit of objectives that could otherwise be achieved through more effective government regulation and community action. Rather than allowing governments and communities to embark on a just transition towards a more sustainable path, CDM may reward government ineptitude and support the very agents that contribute to climate change, while allowing rich countries to continue avoiding the reductions necessary to mitigate climate change.

Source: <www.focusweb.org>

emission through switching or by adopting energy-efficient technologies. It also promotes research on various carbon-neutral technologies and ways to develop and disseminate them. It urges developed countries to help developing countries acquire energy-efficient technologies through their more advanced research and technical capabilities. In one of its articles on climate change, the Kyoto Protocol makes it mandatory for all countries (including the developing countries) to report, on a year-wise basis, on national emission inventories and on measures being adopted to tackle climate change.

The Kyoto Protocol (Articles 4.8 and 4.9 of the FCCC) also restates the principle of protecting vulnerable countries from the potential adverse effects of any of the policies and measures that may be adopted in the fight against climate change, along with the adverse effects of climate change. Examples of vulnerable countries include small island countries, those with areas prone to natural disasters, and countries whose economies are highly dependent on income generated from the production, processing and export, and consumption of fossil fuels, which may get discounted by the stringent regulations of emission cut down.

VOLUNTARY MECHANISM: THE OTHER APPROACH

The voluntary carbon mechanism functions outside of the compliance/regulated Kyoto Protocol mechanism (see Figure 2.3). It enables businesses, governments, non-governmental organizations, and individuals to offset

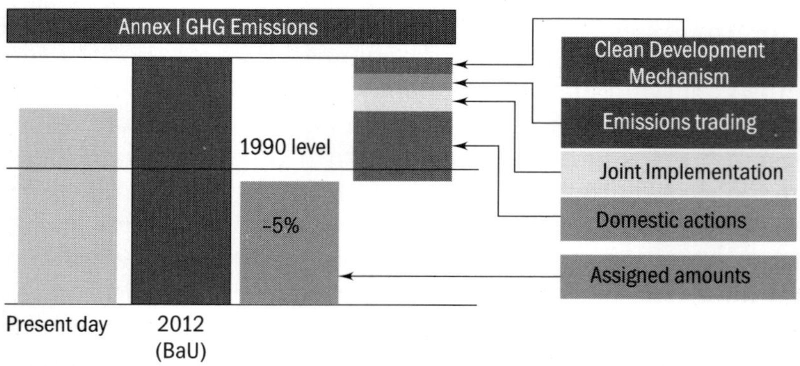

FIGURE 2.3 Flexibility mechanisms of Kyoto Protocol

Source: Yap (2004)

their emissions by purchasing carbon offsets—verified or voluntary emission reductions (VERs)—that were created either through CDM or in the voluntary market. It is noteworthy that about 17% of the offsets sold in the voluntary market in 2006 were sourced from CDM projects (Hamilton, Ricardo, Guy, *et al.* 2007). A carbon offset is:

- A unit earned by someone who, following international standards, has implemented a project that reduces, removes, stores, or avoids GHG emissions that would have occurred otherwise (for example, a wind farm).
- Issued by an authority or a board that follows international standards (1 tonne of CO_2 emission, or CO_2 equivalent, that has been reduced, removed, stored, or avoided is equivalent to one credit earned).
- Bought by someone either for offsetting purposes, in which case the credit is retired (that is, taken out of circulation permanently) to offset their own emissions (where 1 tonne of CO_2 can be offset by one unit/credit), or for investment or speculative purposes (Markit Environmental Registry 2009).

The types of offsetting include the following:

- Removal of CO_2 from the atmosphere and its storage in a "sink", for example, forest regeneration.
- Reduction of CO_2 emissions by using renewable energy sources, for example, wind and solar energy, in place of fossil fuels.
- Capture of GHGs and their alternative use or destruction, for example, methane capture at landfills.
- Emission reduction through energy efficiency, for example, reduction in the amount of fuel or electricity or nitrogen used (Markit Environmental Registry 2009).

In a voluntary market, the demand for carbon offsets is created only by voluntary buyers, such as corporations, institutions, and individuals, whereas in a compliance market, the demand is created by a regulatory instrument. Thus, trading volumes in the compliance market are much larger than those in the voluntary market. In the voluntary market, the demand for carbon offsets is low, their quality standards are not widely established, and they are not fungible in compliance markets. For these reasons, carbon offsets are sold at a lower price in the voluntary market than in the compliance market (CORE 2010).

In the voluntary carbon market, the majority of buyers are private firms that purchase mainly for resale or investment, or to "retire" credits to offset emissions. For some companies or individuals, reduction in their emissions at a small or large scale may not be difficult; however, for some this reduction can be extremely difficult, impossible, or not economically viable. Purchasing credits earned from a verified or voluntary emission reduction project can help offset such emissions. On the other hand, that emission reduction project can be financed by the revenue earned from the sale of those credits. Although the term "voluntary carbon market" is often interchanged with the term "voluntary offsets market", both have the same meaning (Markit Environmental Registry 2009).

In the voluntary market, transaction costs associated with project implementation are fewer. The voluntary carbon mechanism is not guided by any established rules and regulations (unlike under CDM, which is more stringent). Thus, voluntary markets can be easily explored for testing new procedures, methodologies, and technologies, which may later be included in regulatory schemes as compared to CDM or other compliance market projects. Thus, voluntary markets allow for experimentation and innovation. In addition, voluntary markets are suitable for micro-projects, which are too small to warrant the administrative burden of CDM or for projects currently not covered under compliance schemes. However, the main drawback of the voluntary market is the production of low quality VERs, which is a result of poor quality control. To address these quality concerns, several voluntary offset standards have been developed (CORE 2010), which will be discussed in detail in the subsequent chapters.

3
FORESTRY AND CLIMATE CHANGE

H S Gupta

LINK BETWEEN CLIMATE CHANGE AND GHG EMISSIONS

In the forestry and agriculture sectors, three types of greenhouse gases (GHGs), namely, carbon dioxide (CO_2), methane (CH_4), and nitrous oxide (N_2O), are released, taken up, and stored by different sources and sinks. GHGs are emitted to the atmosphere through various agricultural and forestry practices. For example, CO_2 emissions occur when trees are felled to convert land into cropland. The forestry and agriculture sectors are important contributors to climate change. At the same time, they also provide the sink function (that is, forests perform the sequestration function), which helps control climate change. Carbon sequestration is a process in which trees or plants capture atmospheric CO_2 during photosynthesis and store it in their trunks, branches, twigs, leaves, and fruits as cellulose. The process also involves release of oxygen into the air. Figure 3.1 illustrates the process. Figure 3.2 shows the components of the land carbon cycle: how carbon is stored (or sequestered) above the ground by plants, crops, and trees, and below the ground in the soil and roots. Trees and plants take up CO_2 also through roots.

Carbon is released in the atmosphere in many ways. Organic materials get decomposed and release carbon, which is stored in the soil (the amount of carbon in soil is higher than the total amount of carbon in the vegetation and the atmosphere). Animals breathe in oxygen and breathe out CO_2, and carbon and N_2O are released to the soil through animal faeces (FAO 2010).

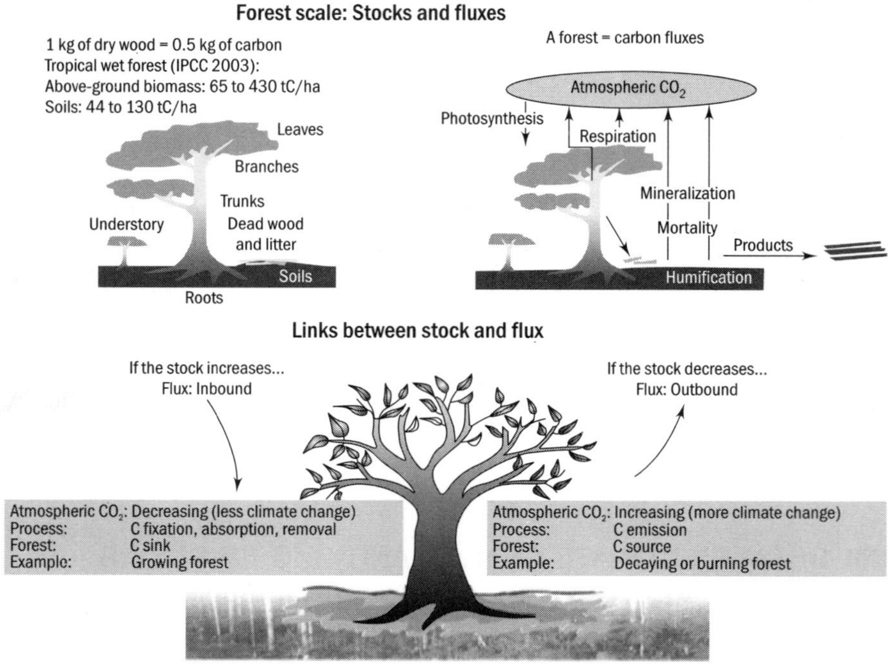

FIGURE 3.1 Carbon sequestration

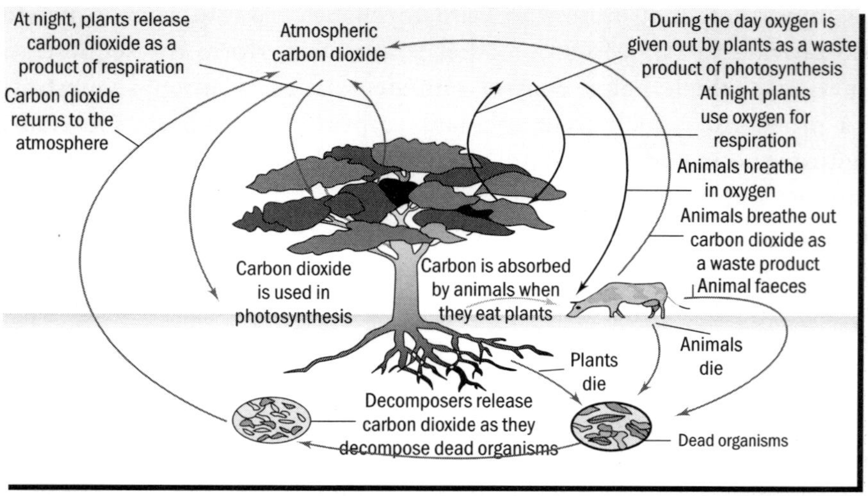

FIGURE 3.2 Carbon cycle at plot level

Source: <www.energex.com.au/switched_on/being_green/being_green_carbon.html>

As already discussed, to slow down climate change impacts, reduction in GHG emissions is urgently needed. We have seen that GHG emissions occur through several agricultural and forestry-related activities. Therefore, reduction in GHGs can be effected by replacing existing practices by new sustainable land management practices (see Box 3.1). Some of these activities are useful in increasing the amount of organic matter and carbon in the soil (by using cover crops) or reducing methane emission (by improving feeding practices). Protection of forests through sustainable forest management can control the release of CO_2; planting new trees can also help in more CO_2 sequestration (FAO 2010).

Following are some of the activities in the agriculture, forestry, and other land use sectors that can help reduce or avoid emissions, or increase GHG removal (Figure 3.3):

- Forestry activities such as afforestation and reforestation, sustainable forest management, agroforestry, avoiding deforestation, and Reducing Emissions from Deforestation and forest Degradation (REDD).
- Silvicultural activities such as plantation and grazing land management, livestock management (improved feeding practices), and manure management.
- Energy activities such as increasing the energy efficiency at household or community level, and sustainable biofuel production.

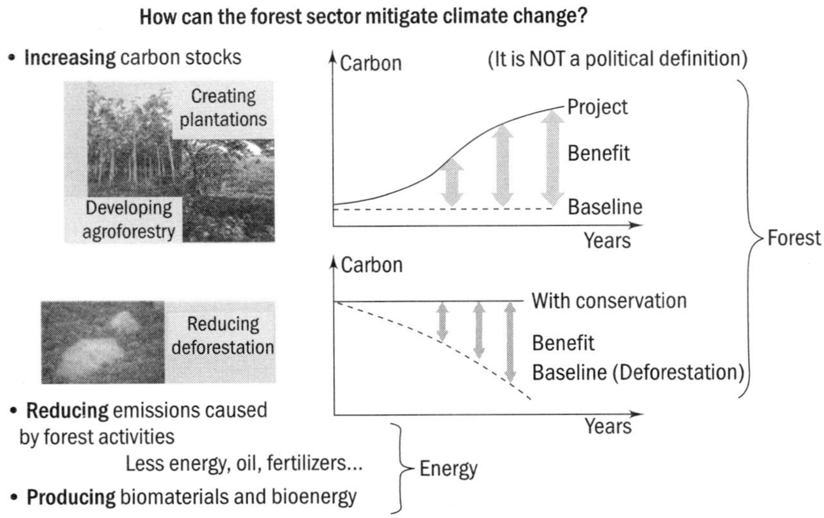

FIGURE 3.3 | Mitigation through forest

- Biodiversity-enhancing activities such as watershed and soil management, and biodiversity conservation (FAO 2010).

CARBON STOCK OF INDIAN FOREST

In India, 63.72 million hectares (Mha) of land is under forest, accounting for 19.39% of the geographic area of the country (FSI 1999). The forests of India are broadly classified into 14 major types, based on climate and altitude. The dominant forest types are tropical moist deciduous and tropical dry deciduous, accounting for 36% and 28% of the total area under forests, respectively (Ravindranath 2006). Table 3.1 gives the carbon stocks in different forest types of India.

In developing countries such as India, the forestry sector plays an important role in the context of climate change and can address the following issues:

- Limiting CO_2 emissions from the land-use change and forestry sector (see Box 3.2).

Table 3.1 Carbon stocks in forest types of India

Forest type	Area ('000 ha)	Standing biomass (MT)	Total carbon stock (MT)	Soil carbon (MT)
Tropical wet evergreen	5,289.3	1,510.72	755.36	702.37
Tropical semi-evergreen	2,575.7	576.44	288.22	442.38
Tropical moist deciduous	23,054.7	4,018.92	2,009.46	1,317.35
Littoral and swamp	383.4	36.28	18.14	11.59
Tropical dry deciduous	18,083.3	731.58	365.79	1,048.65
Tropical thorn	1,573.0	25.10	12.55	69.21
Tropical dry evergreen	1,314.2	2.23	1.12	43.37
Subtropical broad-leaved hill	267.8	11.1	5.56	29.10
Subtropical pine	4,567.5	441.07	220.54	413.04
Subtropical dry evergreen	1,201.0	55.64	27.82	39.63
Montane wet temperate	2,581.9	299.19	149.60	486.30
Himalayan moist temperate	2,242.8	537.52	268.76	314.89
Himalayan dry temperate	30.5	2.10	1.05	2.28
Sub-Alpine and Alpine	2,027.7	124.57	62.29	523.41
Total	64,013.0	8,372.47	4,186.24	5,443.56

MT – million tonnes

Source: Adapted from Ravindranath (2006)

BOX 3.1 ***Extent of forest over the globe***

The total forest area of the world is about 4 billion hectares, which represents nearly 30% of the earth's landmass. Approximately, 56% of these forests are located in tropical and subtropical areas. Forest cover is unevenly distributed. Only seven countries possess about 60% of it, 25 countries around 82%, and 170 countries share the remaining 18%. Planted forests account for approximately 3.8% of the total forest area, or 140 million hectares.

Source: FAO (2007)

- Moderating the potential impact of climate change on people.
- Mitigation opportunities and potential with cost-effectiveness (Ravindranath 2006).

CONTRIBUTIONS OF FORESTRY

Forest is one of the greatest resources given by nature. It is home to the vast and varied biodiversity existing on earth and is valued by communities for its various tangible and intangible products. Many communities are completely or partially dependent on forest for their survival. They directly extract fuelwood, non-timber forest products (NTFPs),[1] and timber from the forest. In addition, forest also indirectly helps provide some vital ecological services such as watershed, rivers, pollination, rain, carbon sequestration, and oxygen. Forest removes carbon from the air and subsequently sequesters it as biomass, thereby acting as a natural carbon sink. This process of storing carbon and then transforming into wood and other parts of plant through photosynthesis is known as carbon sequestration. Trees normally continue to store carbon for years, but this process is relative to their age and species. The capacity of carbon sequestration of a tree reduces as it nears its maturity; that is why forest provides a unique means to mitigate the effects of climate change by absorbing carbon with its own nuances.

ROLE OF FOREST IN FIGHT AGAINST CLIMATE CHANGE

Climate-change-induced effects pose a great threat to human survival; hence, some immediate measures are needed to check growing GHG

[1] Non-timber forest products are the products or services drawn from the forest other than timber, for example, fruits, vegetables, medicinal herbs, resins, essences, and fibres such as bamboo, and rattans.

BOX 3.2 ***Deforestation and climate change***

It is estimated that 1.7 billion tonnes of carbon are released annually due to land-use change. The major portion comes from tropical deforestation, which represents about 20% of current global carbon emissions and is greater than the percentage of carbon emitted by the global transport sector with its intensive use of fossil fuels.

Source: IPCC (2007)

emissions. Among all the alternatives available, forestry can be one of the easiest and cost-effective means to alleviate the adverse impacts of climate change.

The forestry sector can help reduce GHG content primarily through afforestation. At the same time, it can ensure sustainable development. The *Third Assessment Report* of the Intergovernmental Panel on Climate Change (IPCC) also corroborates this point by concluding that the forestry sector has a biophysical mitigation potential of 5380 million tonnes (MT) of CO_2 per year on an average, up to 2050 (Kauppi and Sedjo 2001). In addition, as mentioned earlier, communities benefit from many services derived from forests. Therefore, the growth of the quality and quantity of forest will pave the way for an integrated approach towards development and carbon management. On the contrary, there is no denying that carbon removal through forest is of temporal nature because forest releases carbon into the atmosphere through respiration, decay, and forest fire. But this temporal nature can be avoided if wood is sustainably and effectively collected from matured trees. However, this will be effective only through a strategy that promotes a cycle of planting, growing, and harvesting of trees. This will ensure more carbon sequestration and, at the same time, generation of more wood and NTFPs in the market. According to a recently released GHG inventory report in India, the forestry sector is sequestering 177 MT of GHG emissions.

Climate change can also be controlled through reduction in the deforestation rate. Every year 12.9 Mha of forested area is cleared for various purposes, primarily for converting it into agricultural land (IPCC 2007). By reducing this human-induced unsustainable deforestation, the overall health of forests and their capacity to store carbon can be improved. On a sustainable basis, forest management helps maintain the existing forest with its associate benefits such as watershed protection, biodiversity, livelihood, and environmental security.

In the forestry sector, problems associated with climate change cannot be solved by avoiding timber logging or controlling deforestation, because this will create a shortage of timber in the market. To overcome this shortage, wooden products will be replaced by steel and other metal products, which are more energy consuming; thus, there will be no effective GHG reduction (Gustavsson, Pingoud, and Sathre 2006). Therefore, forests should be managed sustainably by eliminating matured trees and planting new ones in their place. This type of management will ensure constant GHG removal from the atmosphere and also ensure that no leakages would happen due to the maintenance of forest. Therefore, forestry can be of great assistance in combating climate change and also in reducing poverty and enhancing livelihood opportunities by providing environmental benefits along with social co-benefits.

GLOBAL FIGHT AGAINST CLIMATE CHANGE WITH FORESTRY PROJECTS

Forests act as a carbon sink and assist in reducing the carbon content in the atmosphere (as growing trees causes sequestration of CO_2 from the atmosphere). However, sequestration is regarded as temporary storage of CO_2 because cutting and burning of trees again releases the previously stored carbon. There are projects in the energy and other sectors where reduction in carbon emission can be permanently effective. Looking into this non-permanent nature, only two types of forestry projects in land use, land-use change, and forestry (LULUCF) are currently allowed under the Kyoto Protocol. Thus, the current global approach provides lower priority to the inbuilt benefits (livelihood generation, biodiversity conservation, and other social co-benefits) of other forestry projects such as forest conservation and avoiding deforestation.

Despite forest being one of the most trusted allies of human beings, especially for local communities who rely heavily on forest and forest products for their survival, the forestry sector gets differential treatment. Forestry projects are capable of providing an integrated approach to both ecological and environmental restoration, coupled with other co-benefits such as erosion prevention, watershed protection, enhanced biodiversity, and provision of forest resources for local people, subsequently helping the poorest people in the poorest areas. LULUCF projects can help in emission reduction in numerous ways, such as reduction in deforestation, reduction in fire risks, or changes in forest harvesting (for example, changing from conventional logging to reduced impact logging). This shows that in the context of Clean Development Mechanism (CDM), LULUCF projects have an added value (Pearson, Sarah, and Sandra 2006). However, in 2001, in the seventh Conference of the Parties (CoP) in Marrakesh, Morocco, it was decided that at least for the first commitment period, LULUCF projects under the CDM should be limited to afforestation and reforestation activities, which are defined as follows:

- "Afforestation" is the direct human-induced conversion of land that has not been forested for a period of at least 50 years to forested land through planting and seeding, and/or the human-induced promotion of natural seed sources (UNFCCC 2003).
- "Reforestation" is the direct human-induced conversion of non-forested land to forested land through planting and seeding, and/or the human-induced promotion of natural seed sources, on land that was forested but has been converted to non-forested land. For the first commitment period, reforestation activities will be limited to reforestation occurring on those lands that did not contain forest as on 31 December 1989. Figure 3.4 illustrates and distinguishes the concept of afforestation and reforestation.

Land use, land-use change, and forestry include afforestation and reforestation of degraded land, agricultural land, and agroforestry,

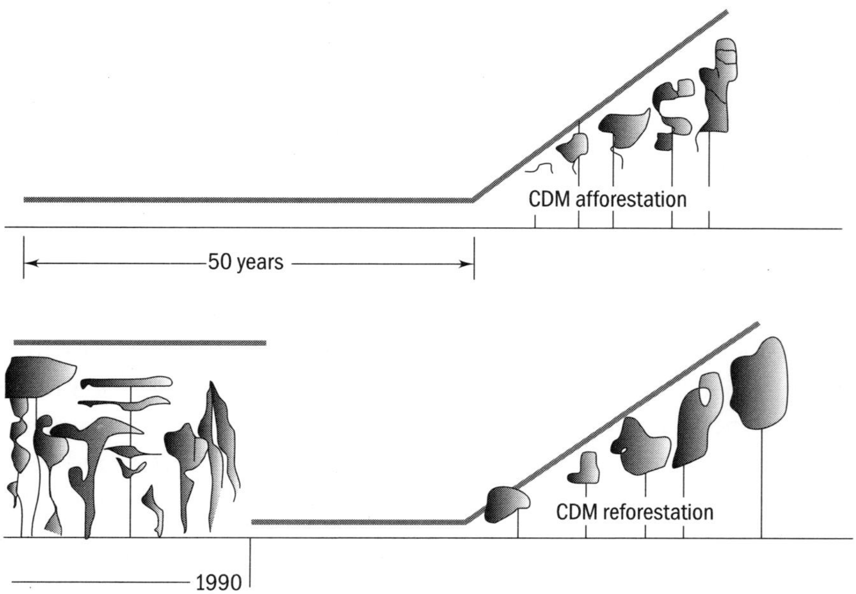

FIGURE 3.4 Afforestation and reforestation, the difference

provided they meet the criteria set up by the CDM Executive Board. Afforestation and reforestation CDM project activities are subject to the specific modalities and procedures of the CDM, which were adopted in the CoP-9 in December 2003. The CDM Executive Board has defined the following threshold values for the definition of forest (Figure 3.5); it has further authorized the host country to decide the adequate definition of the forest within its territory, to segregate the already existing forest area from the project boundary.

- Minimum crown cover: 10%–30%
- Minimum height at maturity of vegetation: 2–5 m
- Minimum area: 0.05–1 ha

The Government of India has adopted the following definition of forest for India:

- Minimum crown cover: 15%
- Minimum height at maturity of vegetation: 2 m
- Minimum area: 0.05 ha

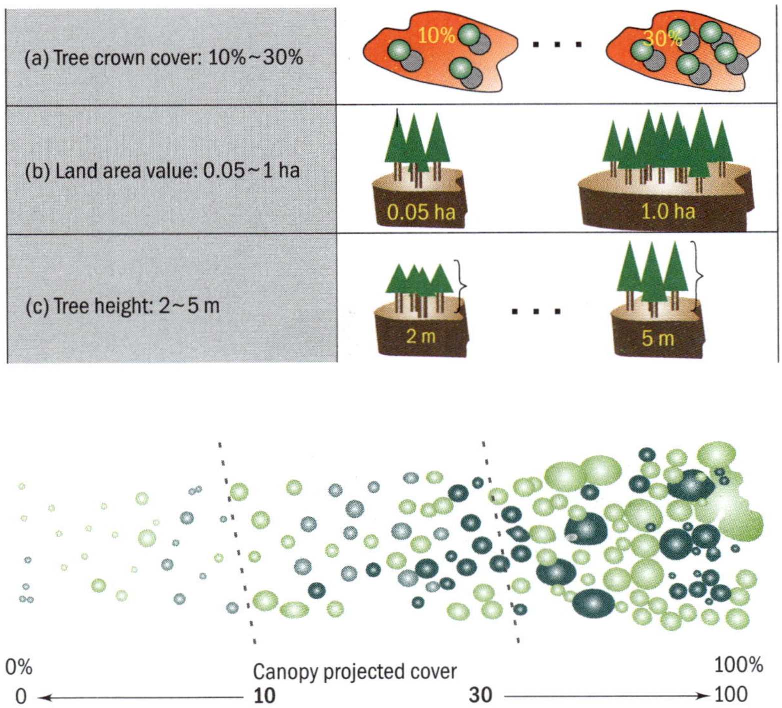

FIGURE 3.5 Canopy cover and forest definition—the parameters

However, the information related to the state of vegetation since 1990 can be difficult to obtain due to the constraints related to the availability of historical land-cover data. Hence, the Executive Board clarifies for this situation that the proof of status of forest in 1990 could also be based on any of the following:

- Aerial photographs or satellite imagery.
- Ground-based surveys.
- A participatory rural appraisal, if the above two options are not available/applicable.

Appendix 2 discusses the procedures to define the eligibility of land for afforestation and reforestation activities. Figure 3.6 highlights the growing importance of forestry projects in the overall portfolio of CDM projects. Similarly, Figure 3.7 depicts the importance of India as a significant player in the carbon project market vis-á-vis China. India and China host most of the projects in Asia, but most certified emission reductions (CERs) are expected from China. The figure also offers a hint to Indian entrepreneurs and government agencies to exploit the vast potential available, using the existing strength of the country.

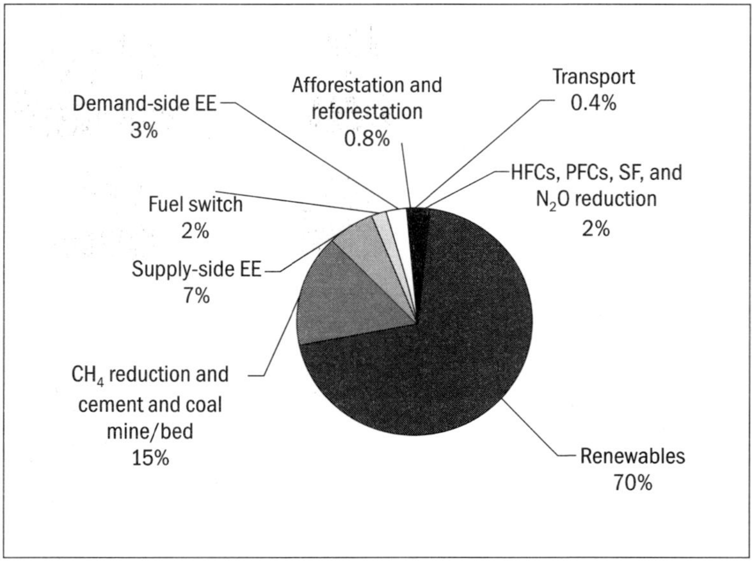

FIGURE 3.6 Percentage of CDM projects in each category

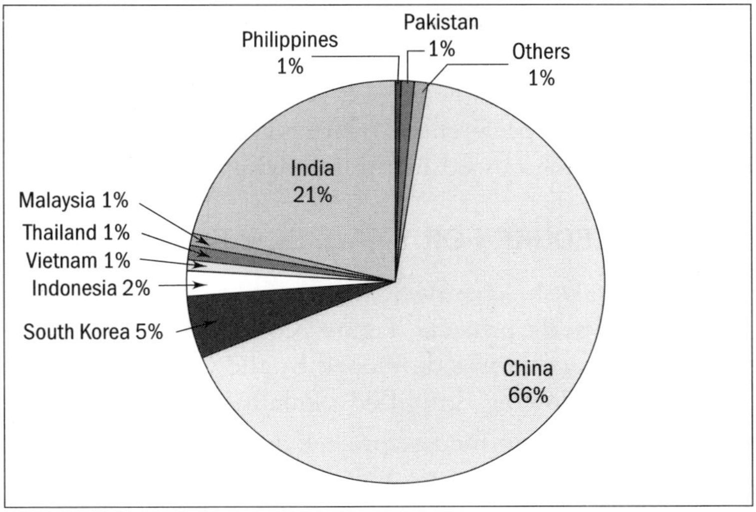

FIGURE 3.7 Percentage of CDM projects by country

CLASSIFICATION OF FORESTRY PROJECTS—ON SCALE

The afforestation and reforestation projects under CDM provide a wider scope for the integration of both environmental sustainability and socio-economic benefits to the communities by availing more livelihood opportunities ensured by additional income generated by the selling of CERs. These projects have also been classified into two categories based on their size:

1. **Small-scale afforestation and reforestation CDM project:** A small-scale afforestation and reforestation project activity is defined as an afforestation or reforestation measure, operation, or action targeting net removal of anthropogenic GHGs of less than 8 kilo tonnes (KT) of CO_2 per year. Generally, small-scale projects are developed or implemented by low-income communities and individuals, as determined by the host party. These projects enjoy certain advantages such as use of simplified modalities and procedures, no adaptation tax, and reduced registration and administration fees. Afforestation and reforestation projects of 16 KT CO_2 represent about 400–800 ha for a typical forest project planting fast-growing species.
2. **Large-scale afforestation and reforestation CDM project:** On the contrary, a large-scale afforestation and reforestation project activity under the CDM is defined as the measures targeting net anthropogenic

GHG removal by sinks greater than 8 KT CO_2 equivalents per year.

This classification of forestry project is done to facilitate the smaller projects by giving them preferential treatment in many stages of project formulation, which is discussed in the following sections.

SIMPLIFIED PROCEDURE FOR SMALL-SCALE PROJECTS

As part of the Marrakesh Accords, it was agreed to establish a simplified procedure for small-scale projects. Figure 3.8 shows a typical cycle of a forestry project. This cycle was developed by the CDM Executive Board and comprises the following simplified modalities and procedures:

- Simplified requirements for the project design document.
- Simplified methodologies to produce the baseline and the monitoring plan.
- Possibility of bundling multiple project activities to form a single project.
- Possibility of using the same Designated Operational Entity (DOE) for validation, verification, and certification.

In addition, projects developed aiming at an annual emission reduction of less than 15,000 tonnes of CO_2 equivalent are exempt from the registration fees levied by the CDM Executive Board. Some

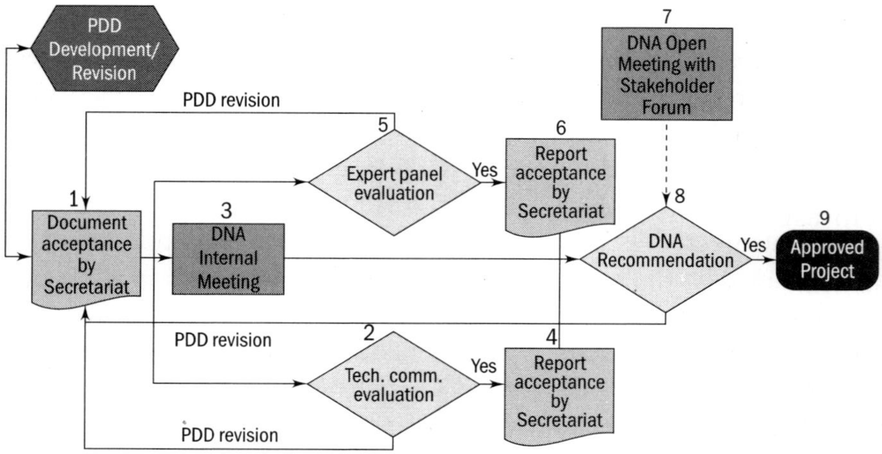

FIGURE 3.8 Project cycle of afforestation and reforestation CDM project

Source: CIFOR (2005b)

characteristics typical of forestry carbon projects have been discussed in the following subsections.

Non-permanence Nature

As discussed earlier, credits issued in CDM projects are known as CERs; however, these CERs are vulnerable to various problems such as fire, deforestation, and natural disasters, which make them temporal in nature. Therefore, in discussions regarding the full inclusion of forestry project into CDM, the major bone of contention remains the risk of losing the sequestered carbon (Pearson, Sarah, and Sandra 2006). Consequently, the parties agreed at CoP-9 that credits arising from afforestation and reforestation projects under the CDM would be temporary, but that they can be reissued or renewed (every five years) after an independent verification certifies that sufficient (equivalent to the issued CERs) carbon is still sequestered within a project to account for all the credits issued. See Box 3.3.

BOX 3.3 ***Latin America: re-finding forestry***

Last few years have seen tremendous growth in the importance of Latin America, particularly in the forest carbon markets. From 2008 to 2009, the number of credits sourced from Latin America more than tripled, as evident from a study of voluntary projects from 10 Latin American and Caribbean countries: Brazil, Peru, Ecuador, Panama, Mexico, Guatemala, Honduras, Argentina, Nicaragua, and Uruguay. However, the majority of Latin American credits originated in Brazil (56%) and Peru (23%).

Brazil's market share mirrors that of 2008, while Peru took the second place from Mexico. Peru's lead results are from a very large Reducing Emissions from Deforestation and forest Degradation (REDD) project as well as several afforestation and reforestation projects and run-of-river hydro-projects. Project developers cite the increased global focus on REDD deployment, as well as national governments' receptiveness and desire to showcase forest carbon projects internationally as catalysts for the region's growth.

In 2009, forest carbon represented 80% of the country's mix. Energy efficiency and fuel switching took most of the remaining market share (14%), while biomass and hydro-projects generated very small volumes.

Source: Hamilton, Sjardin, Peters-Stanley, *et al.* (2010)

Baseline

The baseline scenario is referred to as a business-as-usual scenario signifying the situation that would have happened in the absence of the project. A business-as-usual scenario represents the actual condition where a natural process of carbon sequestration would have taken place and, thus, can be used as a benchmark for calculating additional enhancements in the process of carbon sequestration due to the commencement of project activities.

The CDM Executive Board has already defined some of the methodologies to calculate the baseline scenario for different projects, including afforestation and reforestation projects. It is mandatory for the project proponents to define the baseline scenario in a transparent manner with a conservative approach (Pearson, Sarah, and Sandra 2006).

Additionality

The whole objective of the Kyoto Protocol is to curb the soaring GHG emissions, which trigger climate change. The protocol provides flexibility mechanisms to assist nations and private organizations to achieve that objective. The complexities associated with the mechanism and the need for transparent enhancement of the carbon sinks led to the introduction of the concept of "additionality". This concept was introduced to ensure that emission reduction (carbon sequestration in the case of afforestation and reforestation CDM) should be induced solely by project activities and should not be natural; that is, this reduction should not have happened in the absence of the project.

In 2003, the following definition of "additionality" was agreed on in the CoP-9: "The proposed afforestation or reforestation project activity under the CDM is additional if the actual net GHG removals by sinks are increased above the sum of the changes in carbon stocks in the carbon pools within the project boundary that would have occurred in the absence of the registered CDM afforestation or reforestation project activity...." Therefore, the enhancement of removals by afforestation and reforestation projects must create real, measurable, and long-term benefits related to the mitigation of climate change (Article 12.5b of the Kyoto Protocol), and must be additional to any that would occur in the absence of the certified project activity (Article 12.5c).

Leakage

Although most of the afforestation and reforestation projects successfully sequester carbon within the project area, some projects may alter activities or behaviours elsewhere. These activities may generate unnecessary emissions in and around the project boundary. As a result, the net sequestration of carbon reduces, that is, emissions outside the project boundary increase due to the project activities, thereby negating a portion of the benefits the project aims to deliver. This is called leakage (Pearson, Sarah, and Sandra 2006).

Therefore, leakage is an important concept in the Kyoto Protocol projects with respect to calculating the net carbon sequestered or net emission reduction. For this, the direct or indirect emissions by the project activities are required, which are then discounted from the additional carbon sequestered to obtain the net carbon sequestered.

SCOPE OF FORESTRY PROJECTS IN KYOTO PROTOCOL

So far, LULUCF projects have been excluded from one of the biggest CDM markets, that is, European Union Emissions Trading System, citing the uncertainty related to the non-permanence nature and leakages associated with forestry projects. This restriction has cost a great deal in terms of economic gains and motivation for individuals, communities, and private players in developing forestry projects (Pearson, Sarah, and Sandra 2006). Even under the Kyoto Protocol mechanism, the trading of afforestation and reforestation CDM-related CERs is restricted to only 1% of the total assigned amount per year of the first commitment year. Some additional characteristic features of forestry that need to be considered by any project proponent have been discussed in the next section.

CREDITING PERIOD

The crediting period for an afforestation and reforestation project activity under CDM is the period during which the net anthropogenic GHG removals by sinks are verified and certified by a DOE for issuing CERs. Project participants can choose between the following two options for the length of a crediting period (CIFOR 2005b):

1. **Fixed crediting period** is a crediting period of a maximum of 30 years without the possibility of renewal or extension once the

proposed afforestation and reforestation CDM project activity has been registered.

2. **Renewable crediting period** is a crediting period of a maximum of 20 years, which may be renewed up to two times (maximum 60 years) provided that, for each renewal, the DOE determines that the original project baseline is still valid or has been updated, taking account of new data.

A role has been assigned to forestry by the global regime in the fight against climate change, and that role is too important to be limited to only afforestation and reforestation activities in the Kyoto Protocol. However, even this limited opportunity motivated by voluntary market (discussed in later chapters) provides a wide scope for the development of forestry carbon projects. Thus, forestry projects can prove to be a very effective weapon for the developing countries to fetch some sort of economic equity in the context of climate change. In addition to this, forestry carbon projects with add-on benefits of sustainable development can provide livelihood opportunities for the local and deprived communities. Thus, forestry projects are very important in the context of both climate change and inclusive development, and consequently, local and national governments should promote forestry carbon projects. This chapter is an attempt to highlight the critical measures to be considered while developing a forestry carbon project.

4

VOLUNTARY MECHANISMS FOR FORESTRY CARBON PROJECTS

H S Gupta

Parallel to the markets offered by the regulated Kyoto Protocol regime, a number of compliance markets are also emerging, which are not directly linked to the Kyoto Protocol but are devoted to alternative regional regulatory or regional voluntary compliance efforts. Governmental regulation, energy efficiency, green branding, self-realization, self-discipline, and corporate social/environmental responsibility are the key drivers for these voluntary commitments aimed towards limiting greenhouse gas (GHG) emissions. The markets that are outside the purview of the Kyoto Protocol are broadly categorized as voluntary markets. Some typical examples of premier voluntary markets are Chicago Climate Exchange (CCX), Regional Greenhouse Gas Initiative (RGGI), and the markets of New South Wales (Australia).

Some of these voluntary markets permit the import of project-based carbon credits (from forestry projects) to attain emission reduction targets (Neeff, Lisa, Imme, *et al.* 2007). Such project-based transactions on these markets together amounted to an estimated 17 million carbon credits in 2006 (World Bank 2007). Thus, forestry carbon projects are also eligible for these markets, and the resultant credits are procured by the buyers who volunteer to reduce their GHG emissions or achieve carbon neutral status. These markets are fostered by the private sector, which is obliged to engage in forestry activities by way of self-regulation, or which invests into forestry for reasons related to public relations and public image.

IMPORTANCE OF VOLUNTARY CARBON MARKET

The scope provided by the Kyoto Protocol's flexibility mechanism is too limited for the vast scope of different forestry projects that are capable of generating carbon credits around the world. Moreover, the procedure for registration is also stringent for the forestry projects under the Kyoto Protocol. On the other hand, voluntary markets are driven by individuals, corporations, and other organizations. They provide some respite from the stringent registration procedure, which makes them a more suitable place for the forestry carbon projects. These non-Kyoto markets have the comparative advantage of delivering "charismatic" carbon (Hamilton, Ricardo, Guy, *et al.* 2007). Buyers in the non-Kyoto markets decide voluntarily to purchase carbon credits to use them as offsets for their own emissions, as a growing sense of corporate social and environment responsibility drives the urge to reduce their carbon footprint or even to become "carbon neutral". As a result, this motivation towards green branding has fuelled the voluntary markets with more and more organizations participating and indulging in the trading of carbon.

In the developed and developing countries, there are many such project developers and their number is still growing, who indulge in various types of forestry projects to create carbon offset credits. These project proponents are particularly interested in these new developments because of the Kyoto markets and the rules of the update on markets for forestry offsets within the purview of the Clean Development Mechanism (CDM), which restrict the scope of forestry carbon projects (Neeff, Lisa, Imme, *et al.* 2007). Figure 4.1 gives a broad classification (with examples) of regulatory and voluntary markets for carbon.

SOME IMPORTANT VOLUNTARY COMMITTED MARKETS

Chicago Climate Exchange

Chicago Climate Exchange is the world's first and North America's only voluntary carbon market. It imposes legally binding regulations on its participators, thus making them comply with rule-based GHG emissions reduction. It also creates a regular demand for the carbon credits and, thus, works in favour of project proponents. The exchange started its operations in December 2003.[1] The CCX market uses project-based

[1] Details available at <www.chicagoclimatex.com>

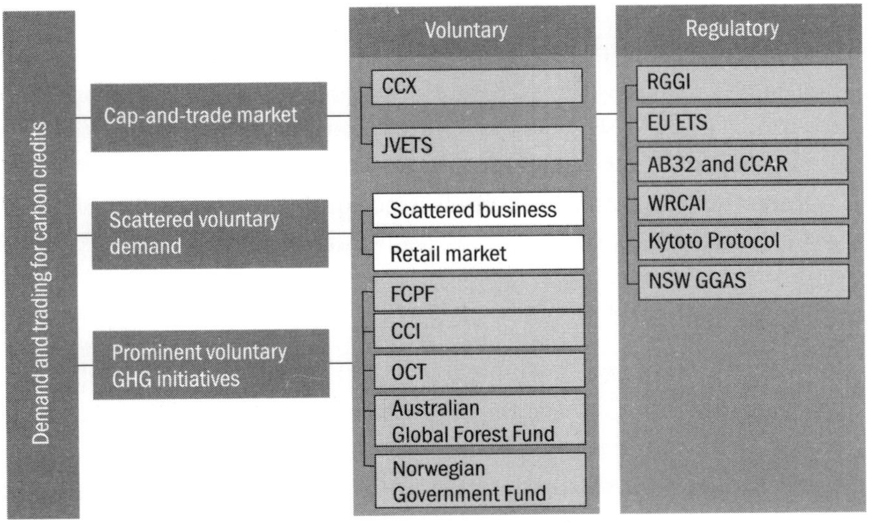

FIGURE 4.1 A schematic diagram of forestry projects

Source: Neeff, Lisa, Imme, *et al.* (2007)

offsets that include forestry. The market is interested in buying forestry-based credits worldwide. Some of the typical features and rules defined by CCX are as follows (Neeff, Lisa, Imme, *et al.* 2007):

- **Volumes and prices:** CCX provides an attractive platform to almost each and every type of carbon trading, but the volume of allowance trading and of other project categories is higher. Even forestry projects have attracted almost 13,986,300 M. tonnes of CO_2 equivalent until 2009.[2] Project-based transactions are on the spot and exclude future offset vintages. CCX seems to provide a lucrative and safe market for carbon offsets from forestry projects and its offered prices have fluctuated within a range of US$2–4 per tonne of CO_2 equivalent.
- **Co-benefits:** The buyer at CCX is not informed about the characteristics of the forestry projects that produce carbon credits. Hence, CCX transactions are project blind. However, trading from unsustainably managed forest is restricted and, thus, the buyer is partially assured of some co-benefits also. To address the issue of non-permanence, the seller is liable for unexpected carbon reversals and therefore has to keep aside 20% of the achieved removals as reserve.

[2] Details available at <http://theccx.com>

- **Additionality:** It is not mandatory for forestry projects to prove the additionality of carbon removals.
- **Eligibility:** Both afforestation and reforestation can earn credits. In addition, CCX also permits "forest conservation activities" to earn credits under some specified conditions, which are as follows:
  - Rare (for example, best-in-class actions)
  - Voluntary (for example, not legally required)
  - Recent
  - Verifiable
  - Properly addresses permanence
- **Conservative:** CCX avoids the creation of perverse incentives that would increase GHG emissions on or off the project site.
- **Certification:** A third-party certification is required by the project developers confirming that they follow the criteria for CCX registration. See Boxes 4.1 and 4.2.

BOX 4.1 ***The voluntary market and the welfare of tribals***

The Climate Trust has signed a contract to purchase offsets from a Lummi Indian Tribe project to acquire previously logged forestland and manage it to permanent old-growth forest. With funding from The Climate Trust, the Lummi Indian Tribe has bought more than 1654 acres of historical tribal forestland near Mount Vernon, Washington, to protect it from logging. The project site is located in the Arlecho Creek watershed and involves a rare mid-elevation cascade mountain forest. A 100-year conservation easement is helping ensure that old-growth forests are developed and preserved.

The project site is being used as an educational laboratory by the Northwest Indian College, which is chartered by the Lummi Indian Tribe and other universities. The laboratory educates both Indian and non-Indian students about measuring the amount of CO_2 sequestered in trees, soil, and other vegetation. It also facilitates research on halting global warming through sequestration of CO_2 in forests.

Source: <www.climatetrust.org/native_northwest.html.com>

BOX 4.2 ***Reforestation of degraded land: an example from India***

Developer: Prakash Industries Ltd

Location: India: multiple sites in Chhattisgarh

Area: 696 acres (282 ha)

Project Type: Reforestation

Rainforest Alliance Evaluation: Validated Compliance with the Climate, Community, and Biodiversity Standards, 1st Ed., "Gold Level": 23 June 2009–22 June 2014 (RA-VAL-CCB-010935)

This reforestation project began in 2002 and will conclude in 2022. Situated on privately owned land located at Rano, Rano II, Pendravan, Saleh, and Lara in the Durg, Rajnandgaon, and Raigarh districts, the project will reforest degraded areas known locally as "Bhata Lands", a name given to lands with poor water holding capacity, little/no vegetation, and with soils prone to wind and water erosion. The project benefits the local community by sponsoring school teachers from nearby villages participating in Durga Puja (a traditional Hindu festival) and other local events such as inter-village cricket matches. The project is a significant source of local employment and encourages women, in particular, to participate in the project. In fact, women form the majority of workers in the Rano and Lara project sites. For these exceptional benefits to the local community, and many other benefits that go above and beyond the basic requirements for validation to the climate, community, and biodiversity standards, the Chhattisgarh project received the standards' prestigious "Gold Level" validation.

Source: Rainforest Alliance (2010)

Oregon Climate Trust

The Oregon Climate Trust was established in 1997. It is a leading non-profit organization that intends to solve the problems of climate change. Its primary activities include investing in high-quality GHG offset projects and promoting carbon offset policy. Eligible projects range from energy efficiency to forestry offsets.[3]

[3] Details available at <www.climatetrust.com>

International Forest Carbon Initiative

Australia's International Forest Carbon Initiative was established in March 2007 with a pledge by the Australian Government to help in reducing deforestation in the Asia–Pacific region by investing US$160 million, in association with other countries that are willing to contribute to this fund. The Australian Global Forest fund promotes investment in forestry-based projects that aim at sequestering GHG emissions. New forest plantation, curbing of deforestation, promotion of sustainable forest management, and capacity building to boost law enforcement to prevent illegal logging are the areas where this fund intends to invest.

New South Wales Greenhouse Gas Abatement Scheme

Established in January 2003, the New South Wales Greenhouse Gas Abatement Scheme (GGAS) aims at reducing emissions associated with the production and use of electricity. With a focus on the forests within New South Wales, GGAS establishes annual statewide GHG reduction targets that are of mandatory nature. Individual electricity retailers need to meet these mandatory benchmarks based on the volume of their electricity market share. The "carbon sequestration rule" under GGAS provides enough scope for forestry projects to be acquired under abatement certificates. Between 2006 and 2012, the annual market volume for the GGAS was estimated to be 15 million abatement certificates.[4]

Regional Greenhouse Gas Initiative

The RGGI was officially established in 2003. It is a regional carbon market spread across several north-east and mid-Atlantic American states. To reduce GHG emissions, the RGGI participating states are using a market-based cap-and-trade programme with a market-based emissions trading system. Under the proposed programme, participating states require to reduce CO_2 generated by their electric power generators. This system is also employing offsets (GHG emissions reduction or sequestration projects outside the electricity sector) to help companies meet their compliance obligations.[5]

[4] Details available at <www.greenhousegas.nsw.gov.au>

[5] Details available at <www.rggi.org>

California Climate Action Registry

California's Global Warming Solutions Act (AB32) was implemented to cap GHG emissions across the state. The primary activities of the California Climate Action Registry (CCAR) include documenting emission reductions. It also serves as a voluntary GHG registry that protects, encourages, and promotes GHG emissions reduction through early actions. Since 2001, the registry has been helping companies and organizations that operate in this state to establish GHG emissions baselines against which any future emissions reduction requirements may be applied. Although the registry is not directly involved in offset projects (being only a reporting mechanism), its members can report project-based offsets that they have purchased to counteract their emissions. Even the developing countries can avail this additional reporting opportunity for their forestry projects. CCAR's first offset protocol is for forestry; hence, it is obvious that the forest sector will have majority of the offset credits.

Norway Government Fund

Norway has announced that it wants to become carbon neutral by 2030. To achieve this, it is all set to acquire carbon offsets from forestry projects across the world. By 2020, Norway aims to attain a target of 40% below the 1990 level and intends to acquire a carbon neutral status within another 10 years. Recently, the Norwegian Parliament has authorized the Ministry of Finance (MoF) to purchase emission credits based on the flexible mechanisms of the Kyoto Protocol. During 2008–12, the MoF planned to buy approximately 27 million tonnes for delivery from both forestry projects and joint implementation. The MoF has invited offers from projects in all stages of the investment process.[6] To promote the development of the carbon market, it has particularly focused on projects from those countries that currently have few or no registered projects. Due to its significant carbon market volume, it can accept relatively high performance risk and at the same time offer the prospective supplier of credits a fair market price. For the 2008–12 period, the minimum volume for each contract was 50,000 tonnes of CO_2 equivalent (MoF 2009).

6 Details available at <www.treehugger.com>

OVER-THE-COUNTER MARKET

There are many transactions related to forestry credits that remain unnoticed because they remain away from the purview of exchanges. These transactions are scattered because there is no structure from a central funding mechanism, an exchange, or a registry. Others have used a related category and termed it "over-the-counter" market to include all transactions on the voluntary market that are not facilitated by exchanges (Hamilton, Ricardo, Guy, *et al.* 2007). These transactions are instigated by individual or organizational motivations to fetch a carbon neutral status. Also these transactions are done between seller and buyer only through an agreement, remaining away from exchanges.

Overall, a voluntary market, along with a regulated market, provides a greater scope for the forestry carbon projects. Forestry projects that include afforestation, reforestation, and even conservation forestry or other forestry activities can earn carbon credits through these well-established, non-regulated markets. This voluntary step to curb emission is promoting the revival of traditional forestry as well as the innovative forestry that integrates sustainable development, with the ultimate objective of environmental conservation. Although these voluntary schemes and markets lack adequate information on the rules and technicalities of trading, their intention to promote forestry projects is notable. An exception is the CCX, which has well-defined mechanisms and procedures for the registration and trading of forestry projects. Soon other markets are expected to follow the way of CCX to pave the way for an ideal market situation for forestry projects.

5

Formulation of Forestry Carbon Projects for CDM or Voluntary Markets

H S Gupta

Forestry carbon projects have the potential to tap market opportunity conforming to either any regulatory regime (including the Kyoto Protocol) or any voluntary market mechanism. Carbon projects meant for these two major markets have certain but different intricacies, which must be given adequate importance by project formulators. The specifics related to these major markets have been discussed in detail in the following sections. Box 5.1 discusses the generic elements that are necessary to be taken care of for the development of a forestry carbon project.

FORESTRY PROJECTS UNDER CLEAN DEVELOPMENT MECHANISM

The Clean Development Mechanism (CDM) under the Kyoto Protocol allows limited participation of forestry projects. Only afforestation and reforestation projects are allowed, as discussed earlier. The Kyoto Protocol proposes a particular format for the documentation of afforestation and reforestation projects. The project proponent must have a detailed understanding to develop an accurate procedure for the project. This understanding also helps in easy registration of the project under the CDM Executive Board. The forestry project involves the following stages.

BOX 5.1 ***Generic elements in the design and development of a forestry carbon project***

Before starting a carbon project, it is important to keep several points in mind. Designing and developing a carbon project takes a long time and requires a lot of technical expertise and considerable financial resources for the initial set-up. There are 10 steps to consider for developing a carbon project. The first five are simple checks for determining whether the project idea is feasible. The last five steps have to be taken up with a project developer who thinks that the project is viable.

1. **Type and scope of the project:** A clear idea of which type of project should be developed, for example, afforestation, reforestation, improved farming techniques (soil carbon sequestration), or renewable energy project, and where the project should be implemented.
2. **Resources check:** One needs to be aware that a significant amount of time and money need to be invested to develop a carbon project. Analyse why it would be attractive to engage in undertaking a carbon sequestration project and what are the driving motivations?
3. **Project group:** Farmers/villagers need to be identified who want to participate and have land or forest that can qualify for the project type determined in step 1. The project boundary (geographically) has to be established. The project area needs to be big enough to generate enough emission reductions to qualify for a carbon project; for a Reducing Emissions from Deforestation and forest Degradation (REDD) project, the minimum project area is around 30,000–40,000 ha, and for an afforestation and reforestation project, it is around 10,000 ha. Small-scale afforestation and reforestation projects under the Clean Development Mechanism (CDM) must result in greenhouse gas (GHG) removals of less than 16,000 tonnes of CO_2 per year. In addition, clear land use and tenure rights are essential.
4. **Institutional back-up:** To organize, aggregate, and represent farmers, an institution is required, such as a community-based organization (CBO), farmer cooperative, or non-governmental organization (NGO), which is trusted by the project participants. It should have a robust and transparent institutional set-up. In addition, it is of advantage if the institution has some expertise on carbon project development, carbon measurements, and accounting and business plan development.

Contd...

Box 5.1 Contd...

5. **Funding:** Develop a business plan that takes into account all costs and benefits of the project. Ensure sufficient funding for the initial set-up of the project. With the information gathered in the first five steps, a project idea note (PIN) should be developed, which can be used further.
6. **Identification of project developer:** In collaboration with the institution, a project developer has to be selected who can assist with the formulation of the project. The project developer is responsible for preparing it for the market. The developer can be a back-up institution (step 4), if it has sufficient experience, or a specialized project developer company (for example, Ecosecurities, Ecopositive, EC Carbon, Terra Global Capital, and Carbon Neutral Company), or the World Bank Carbon Finance Unit.
7. **Further steps with project developer:** From the different available standards (see Annex 2), the appropriate one has to be selected, market demand assessed, costs and revenues calculated, and a commercialization strategy developed. The project developed should start to select potential credit purchasers.
8. **Project planning/development:** The baseline and methodology need to be selected. Projects must use approved methodologies to calculate emission reductions. The project's chance of being registered and the likelihood of more rapid project preparation increase on using approved methodologies. Developing new methodologies can be resource and time intensive and may not be justified for smaller projects. Assess additionality, leakage, and permanence, and estimate the full GHG inventory of the emissions and uptake of the project. All this information will be assembled in a carbon project document.
9. **Validation:** The project developer determines a third-party certifier (accredited by a specific carbon standard) who will review the carbon project document. It is important for the project to be validated to ensure the transparency of the project design.
10. **Registration:** The voluntary emission reductions (VERs) of the validated project are kept in a registry on behalf of the owner until they are bought.

Source: FAO (2010)

Project Identification

In the project identification stage, project proponent should identify the potential site, the estimation of labour with cost, the possible assumption of additionality, and its baseline scenario. The project should meet all the criteria set up by the host country regarding sustainable development. This is an important prerequisite for the development of forestry projects under CDM, along with an accurate assumption for the afforestation and reforestation activity (according to the country definition) that the project is going to take up. In addition, it should also be estimated whether the project is a small- or a large-scale one, based on the potential delivery of certified emission reductions (CERs).

Project Idea Note

The project idea note (PIN) comprises a short detail of the project. Hence, it should be prepared as per the requirement of the buyers and the markets. For preparing an ideal PIN, project developers should take care of the following issues (adapted from Pearson, Sarah, and Sandra 2006):

- Determine the primary goals of the project according to the requirements of all stakeholders. Goals can be financially, environmentally, and development oriented.
- Based on these goals, identify a cost-effective and suitable afforestation and reforestation activity for the project.
- Identify a boundary in which the project activities will be limited and then prepare a PIN of around four to five pages.

Project Concept Note

Another most important aspect of an afforestation and reforestation project is determining the concept of additionality. In a concept note, one should first clarify the issue of additionality with the required methodology. One should also attempt to elaborate the sustainable development indicators (including social, environmental, technological, and economic well-being) because afforestation and reforestation projects with both environment and social co-benefits are preferred by buyers and investors.

In this note, the baseline scenario, methodology, project boundaries, scope, and the life span of the project should also be discussed in detail. In addition, financial indicators (flow of additional investment, internal rate of return of the project with and without CERs, agreements with stakeholders, and availability of funding), technological feasibility, risk analysis, and credentials with government endorsement should be clarified here.

Development of Project

Project Design Document

The final step of project development is preparing the project design document (PDD) and submitting it for validation. The PDD format and a real example from India are included in Appendices 3 and 10 for the convenience of the reader who would like to design a project.

Baseline Design

The baseline may be designed by the proponent or adapted to fit the existing methodology. Baseline study can be done with an appropriate and approved methodology, as approved by the United Nations Framework Convention on Climate Change (UNFCCC) (see Annexes 7, 7-A, and 7-B for reference). It should consider the actual scenario that would have happened in the absence of the project.

Local Stakeholder Consultations

Consultative meetings should be held with the involvement of all stakeholders at each and every stage of project development. Designated National Authority (DNA)[1] representatives should be invited to these meetings as observers, if possible. The PDD should explain the process by which stakeholders were engaged in the project. The participation of the stakeholders should be recorded with their comments, and the PDD must include a compiled summary of the comments.

Validation

Validation is the process of independent evaluation of a proposed afforestation and reforestation project activity under the CDM by the

[1] The Designated National Authority (DNA) is established by the national government. The Conference of the Parties (CoP) ensures that projects are consistent with national policies with the help of DNA. It may comprise various types of agencies or institutions, depending on the need of the parties. In India, the central government has constituted the National CDM Authority under the Ministry of Environment and Forests.

Designated Operational Entity (DOE)[2] against the requirements set out in decision 19 of the ninth meeting of the Conference of the Parties (CoP) and other relevant decisions, on the basis of the PDD. During the validation of a project, the DOE ascertains the legality and accuracy of the assessments that are made through interview, site analysis, and inspection. The validation process normally takes around four weeks.

Registration

After successful validation, the DOE submits the project to the CDM Executive Board[3] for the final approval. For the verification, certification, and issuance of temporary CERs (tCERs) or long-term CERs (lCERs) to an afforestation and reforestation CDM project activity, "registration" is required. Normally, this process does not take more than eight weeks. Registration is the formal acceptance by the Executive Board of a validated project as a CDM project activity. Registration is deemed final eight weeks after the date of receipt of request for registration unless there is a request for review.

Monitoring

Monitoring refers to the maintenance of the project site for hassle-free delivery of CERs and protection of the site from potential damages. It is also referred to as a collection and archiving of all relevant data mandatory for estimating or measuring the net greenhouse gas (GHG) removals by sinks, which are created by the project during the crediting period, pertaining to all the activities carried out by the project participants.

Verification

Verification is the periodic independent review and ex post facto determination by the DOE of the net anthropogenic GHG removals

---

[2] The Designated Operational Entity (DOE) is designated by the CoP and MoP. Based on the recommendation of the Executive Board, the DOE can validate the proposed Clean Development Mechanism (CDM) project activities as well as verify and certify net anthropogenic greenhouse gas (GHG) removals by sinks. Generally, a single DOE does not perform all these functions on the same afforestation and reforestation CDM project activity; however, upon request, the Executive Board may allow this.

[3] Formed by the CoP, the Executive Board reports to the CoP and MoP. Its main role is to supervise the implementation of CDM. The Executive Board can recommend the revision and amendment of modalities and procedures of CDM implementation, approve methodological issues, and recommend the accreditation of a DOE to CoP and MoP.

by sinks achieved by an afforestation and reforestation project activity under the CDM since the inception of the project. For small-scale CDM projects, the DOE agency validating the project is also allowed to execute the monitoring plan and project verification (CIFOR 2005). The verification and certification of CERs must be conducted by a DOE different from the one that carried out the validation.

Certification

Certification of the project is done by the CDM Executive Board after the DOE provides a written confirmation that an afforestation and reforestation project activity under the CDM has achieved the net anthropogenic GHG removals by sinks since the start of the project as verified. The duration of this process is 15 days. Subsequently, the CDM Executive Board issues a certification followed by the issuance of CERs to the project applicant entities. This step is repeated after every round of submission of verification report.

Issuance of CERs

The issuance of lCERs or tCERs refers to the instruction by the Executive Board to the CDM Registry Administrator to issue a specified quantity of lCERs or tCERs for an afforestation and reforestation CDM project activity into the pending account of the Executive Board in the CDM Registry. The time frame for developing a carbon project that needs to be taken into account is approximately 12 months until the project is validated and 1.5 months for the registration.

Afforestation and Reforestation Methodology and Project Evolution

The registered projects under the afforestation and reforestation sector are quite few compared to the overall CDM (Table 5.1). Some of the reasons for this are as follows:

- For the projects in this sector, the modalities and procedures were created later than those in other sectors.
- The methodology approval process is strict and time consuming.
- Afforestation and reforestation methodologies are very complex to apply.[4]

---

4 Details available at <www.carbonfinance.org>

Table 5.1 Comparing the evolution of methodology and projects: afforestation and reforestation versus other sectoral projects

	2001	2002	2003	2004	2005	2006	2007	2008	2009	2010	Total
Designs on methodologies and procedures											
Other sectors	x										
Afforestation and reforestation CDM			x								
Approved methodologies											
Other sectors			10	22	29	22	21	11	17	17	149
Afforestation and reforestation CDM					1	4	8	3	1	1	18
Registered projects											
Other sectors				1	62	408	427	431	673	508	2,510
Afforestation and reforestation CDM						1	0	0	10	6	17

Source: Adapted from www.carbonfinance. org

CDM Regulations

Compliance with the afforestation and reforestation CDM requirements is quite challenging and should always be kept in mind at different stages.[5] During preparation, the following challenges are notable:

- Difficulty in selecting a suitable methodology due to unclear and overlapping applicability conditions.
- Unavailability of information required for selecting a baseline and demonstrating additionality.

During validation, the challenge to consider is the delay in validation of projects because of multiple revisions to methodologies and low capacity of projects to comprehend regulation and provide supportive documentation.

During verification, problems occur when a project deviates from the PDD, resulting into changes in the monitoring plan. The redrafted plan has to be submitted again for the CDM Executive Board approval, which delays verification and credit issuance. See Box 5.2. The reader would also be interested in detailed designer manual of carbon forestry project, developed by UNFCC (Annexure 13).

DEVELOPMENT OF FORESTRY CARBON PROJECT UNDER VOLUNTARY MECHANISM

Forestry projects other than afforestation and reforestation are not eligible for the globally regulated CDM mechanism under the Kyoto Protocol. However, there could be a variety of forestry projects that can sequester carbon, including forest conservation, reduced deforestation, and degradation projects. Voluntary markets are being established to recognize these kinds of projects and thus provide another scope for earning carbon credits outside the purview of CDM. However, for the documentation of forestry projects to be traded in voluntary markets, these voluntary carbon mechanisms are fragmented with complex supply chains and numerous emerging standards, which provide relevant information required during project development. Moreover, historical data being limited, assessment of the benefits and drawbacks of each standard provider becomes difficult. Only some such offset projects are independently verified to agreed-upon standards, but not all. These standards are numerous and overlapping, but as technology and

---

5 Details available at <www.carbonfinance.org>

BOX 5.2 *Limitations of CDM land eligibility rules*

The project developers are struggling to comply with CDM "land eligibility rules", and they are seeking some changes, some of which are as follows:

- The afforestation and reforestation CDM "land eligibility rules" are quite inflexible to accommodate land-use decisions.
- Developers of multi-stakeholder projects have to struggle a lot for getting farmers to sign contracts that are required for validation.
- Demonstration of land eligibility is a challenging job in the absence of official records and satellite imagery, as well as the low capability of projects to interpret data and information.
- The "1990 rule" excludes areas with significant potential for afforestation and reforestation, and results in scattered planting plots. This negatively affects the social, ecological, and financial aspects of projects.
- By excluding areas that are temporarily stocked with carbon, the "land eligibility rules" make projects on agricultural lands in the tropics challenging.

Source: <www.carbonfinance.org>

attention increases scrutiny, ignoring them may lead to risks of ineffective reductions, unintended but adverse consequences, and even accusations of green washing (Hamilton, Ricardo, Guy, *et al.* 2007).

Box 5.3 lists how agencies are working towards new markets for ecosystem services. Table 5.2 lists the existing different standards for voluntary markets.

These carbon standards primarily assist in project development by providing guidance. Secondly, they help in certification and registration along with some standards that help in providing market linkages to the project proponents. Therefore, project developers should analyse these standards well before the conception of the project and its implementation. Project proponents should take cue from a particular standard, and then they should build projects seeking guidance from the same standard. This will help in ensuring the success of a project and earning revenue from carbon credits.[6]

[6] Details available at <www. caaltd.org>

Table 5.2 Major voluntary market standards

Name of standard	Description	Focus on environment and social benefits	Reporting/ registration	Product label?	Includes LULUCF methodology?	Geographical reach	Start date
Gold Standard (www.cdmgoldstandard.org)	Certification for offset projects and carbon credits	Yes	VER registry in development	Yes	No, energy projects only	International	First project validated in 2006, first credits verified in 2007
The VCS (www.v-c-s.org)	Certification for offset projects and carbon credits	No	Use Bank of New York; other registry				
TBD	Yes	Yes	International	2007			
Climate, Community, and Biodiversity Standards (www.climate-standards.org)	Certification for offset projects	Yes	Projects on website	Yes	Only LULUCF	International	First project certified in 2007
CCX (www.chicagoclimatex.com)	Internal system for CCX offset projects and CCX carbon credits	No	Registry incorporated with trading platform	No	Yes	International	2003
Plan Vivo (www.planvivo.org)	Methodology and certification for offset projects and carbon credits	Yes	No	No	Community-based agroforestry	International	2000
WBCSD/WRI protocol	Guidelines for projects and corporate GHG accounting	No	Does not include registry	No	Protocol created for LULUCF	International	2001

Contd...

Table 5.2 *Contd...*

Name of standard	Description	Focus on environment and social benefits	Reporting/ registration	Product label?	Includes LULUCF methodology?	Geographical reach	Start date
VER+ (www.global-greenhouse-warming.com)	Certification for offset projects, carbon credits, and carbon-neutral products	No	TÜV SÜV				
Blue Registry	Yes	Yes, joint implementation or CDM methodology	International	2007			
ISO 14064 (www.iso.org)	Certification for emissions reporting offset projects, carbon credits	No	No	No	Yes	International	Methodology released in 2006
VOS (www.global-greenhouse-warming.com)	Certification for offset projects and carbon credits	No	TBD	TBD	Follow CDM or joint implementation methodology	International	TBD

CCX – Chicago Climate Exchange; CDM – Clean Development Mechanism; GHG – greenhouse gas; LULUCF – land use, land-use change, and forestry; TBD – to be determined; VER – verified or voluntary emission reduction; VOS – voluntary offset standard; WBCSD – World Business Council for Sustainable Development; WRI – World Resource Institute

Source: Adapted from Hamilton, Ricardo, Guy, *et al.* (2007); Hamilton, Ryan, Emma, *et al.* (2008)

BOX 5.3 ***The growing realization for carbon from forest***

Forestland owners and managers have witnessed the development of markets for many carbon services; however, market-based solutions to compensate forestland owners for these benefits have been limited. Fortunately, new markets for ecosystem services such as carbon sequestration are seeing increased activity and growth. Currently, the most mature of these new emerging markets is carbon.

Agencies like Forecon Inc. have worked intensively in the area of ecosystem services over the past years, building knowledge and influencing policy as these markets develop domestically. As the manager of numerous landowners of all types, it is intimately familiar with the pressures being placed on timberlands. Taxation, regulation, certification, and urban expansion are but a few of the threats to timber resource today. In recognizing the increasing pressures on forestland owners, it is imperative that all market-based opportunities for managed forests be researched and, where appropriate, made available to those that could benefit.

The current domestic carbon markets are rapidly evolving. Rising values for carbon financial instruments, along with favourable policy and accounting rules for forestry projects, are creating exciting new opportunities. It has been identified that these new emerging markets are potential return drivers for some forestland owners who are willing to make commitments to a positive net flow of carbon over time.

Some agencies, like Forecon Inc., can assist considerably the impacts of participation in domestic carbon markets. They provide an overview of the services available, the forest project requirements, the project development process, and the project questionnaire. Thus a client gets help by getting this information, broadening his or her understanding of carbon accounting, and marketing for forestry projects.

The project developer may like to proceed with investigating how the chosen forestry project can participate. All related questions about the project can then be specifically addressed, and the appropriate advice can be provided for the project's success.

Source: Adapted from Director of Ecosystem Services, Forecon Inc.

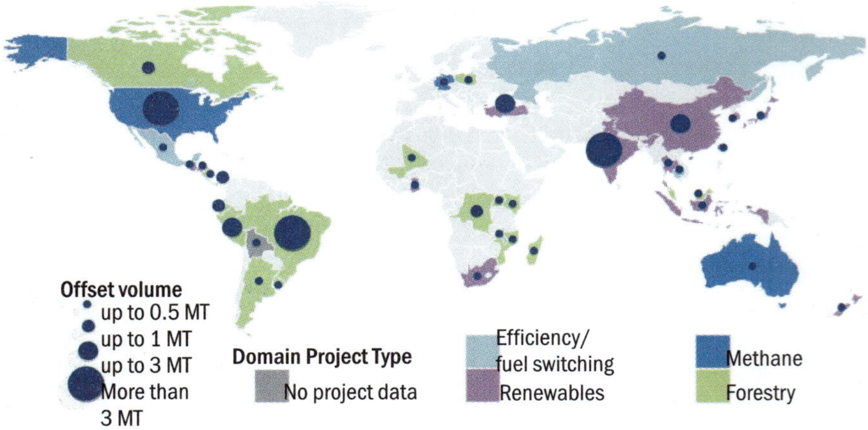

FIGURE 5.1 || Transaction volume and project type by location

Source: Hamilton, Sjardin, Peters-Stanley, *et al.* (2010)

Trends in regional distribution of sales can be attributed to factors on the supply and demand fronts. On the supply side, factors such as country-level policies, risk, infrastructure, availability of project types, and environmental attributes influence the choice of project location (Figure 5.1). At the same time, many buyers tend to have specific preferences with respect to the project location.[7]

CONCLUSION

Project formulation always plays an important role in the success of a forestry carbon project. The rules and protocols for projects defined in CDM are much specified, whereas in the case of voluntary mechanism, these are scattered and variable, as the whole voluntary market is scattered. CDM defines various methodologies and frameworks to guide project proponents; on the contrary, voluntary market defines standards individually and specific to their own market. Hence, care needs to be taken when any project developer frames a project in line with voluntary market and intends to trade carbon credits in the scattered voluntary market. For instance, any project proponents who aim to trade carbon credits generated by its project in the Chicago Climate Exchange (CCX) have to comply with the standards specified by CCX.

[7] Details available at <www.bnef.com>

6

ECONOMICS OF FORESTRY CARBON PROJECTS

M Verma and H S Gupta

Trees growing in forests and farms, being one of the world's largest sinks of carbon, play an important role in global climate change mitigation. Global emissions can be controlled significantly by afforestation alone. Many more unexploited opportunities exist for afforestation, reforestation, and other forestry activities in the developing world.

ECONOMIC ASPECT OF FORESTRY PROJECTS

Concerns over climate change have led to the establishment of markets for the trading of greenhouse gas (GHG) reductions mainly after the introduction of the Kyoto Protocol. Since then carbon trading has grown significantly, reaching a value of $10 billion in 2005 (Capoor and Ambrosi 2006). Major transactions of carbon credits have occurred within the European Union Emissions Trading Scheme, which however restricts trading of forest-based carbon offsets. Such restrictions have so far limited the market for forestry carbon projects. However, forest carbon sink projects are still very beneficial and important (as specified in Article 12 of the Kyoto Protocol) for helping the developing countries achieve sustainable development while contributing to meet the emission-reduction commitments by Annex I countries. The markets of certified emission reductions (CERs) generated by forestry projects are totally demand driven in contrast to energy projects. Yet there exists a substantial demand for forestry projects, and to cater to this requirement, forestry carbon projects can be promoted by the developing nations in a big way. This, however, is constrained by higher risks and costs involved in forestry carbon projects, which generally differ from other carbon projects in terms of cost per unit of carbon emissions avoided or carbon sequestered, and other environmental and social benefits provided (Cacho 2006).

Thus, the supply of forestry CERs or voluntary emission reductions (VERs) depends significantly on the cost-effectiveness and risk-mitigation measures taken in the context of forestry. This chapter attempts to discuss the economics (rather microeconomics) of carbon forestry projects and the various costs associated with them, which are considered important from the point of economic feasibility.

Global-level studies of the economics of climate change mitigation indicate that afforestation and avoided deforestation are among the most attractive investments for reducing net GHG emissions (total emissions less total sequestration) (Swallow, van Noordwijk, Dewi, *et al.* 2007). They also offer direct carbon credit benefits to producers.

However, the concerns of "flooding" the market for carbon credits (mainly of forests) have previously restricted the entry of these credits (in full capacity) in the global rules, despite being less costly than investment in clean energy. A modified political platform now exists for considerably stronger emission reductions, under the Clean Development Mechanism (CDM) and other voluntary mechanisms (Swallow, van Noordwijk, Dewi, *et al.* 2007).

An analysis of the economic returns associated with the land-use transitions (measured in terms of discounted net present value) confirmed the existence of clear economic rationale for almost all land-use transitions occurring at various test sites, as illustrated in Box 6.1. Therefore, almost every land-use transition has been economically rational from the perspective of private land users. The land-use transitions occurred as a result of the following factors: market incentives to harvest and sell timber (see Box 6.2), market opportunities for new cash crops, lack of incentives required to maintain the value of standing carbon, and high interest rates in local financial markets (Swallow, van Noordwijk, Dewi, *et al.* 2007).

A study observed that a large-scale amount of land-use changes for carbon sequestering has also increased the net returns to farmers. This implies that incentives for afforestation and reforestation may encourage further land-use changes, which may increase farmers' income along with

BOX 6.1 ***Economics decides: opportunity costs of CO_2 emissions from land-use change***

A study has helped in understanding how to undermine the strong economic motivations behind deforestation. The results from Peru clearly show that more the people discount the future, the greater the incentive to deforest and the lower the incentive to invest in reforestation. This logic holds true at both the private and public levels. Farmers with poor access to credit and insurance and low security of property rights will tend to have higher discount rates. Governments concerned about short-term election prospects will also tend to disregard long-term projects such as forest protection and sustainable forest management.

Given the market incentives for extraction of timber resources, there will always be a need for enforcement of forest boundaries as part of landscape solutions. Situations such as those found in parts of Indonesia, where most deforestation occurs illegally within and around the national park, undermine the credibility of plans for implementation of the Reducing Emissions from Deforestation and forest Degradation (REDD).

There are considerable institutional challenges for carbon conservation incentive mechanisms. A few of these challenges have been highlighted by the results of the study. First, there is a close relationship between market access and opportunity costs of carbon; poor market access translates into low opportunity costs of CO_2 emissions from deforestation. But poor market access is likely to be associated with weak institutional development and high costs for administering and monitoring carbon conservation contracts. Further, poor market access is likely to prompt farmers to focus on minimizing risk and subsistence food production, likely making them less responsive to market incentives for carbon conservation.

Source: Swallow, van Noordwijk, Dewi, *et al.* (2007)

sequestering carbon. The study also showed that there is an important economic opportunity in establishing multi-strata agroforestry systems on degraded lands, where farmers integrate a range of trees into their farming systems (Swallow, van Noordwijk, Dewi, *et al.* 2007).

BOX 6.2 ***Economic dimension of forest***

In 2003, the international trade in sawn wood, pulp, and paper and boards amounted to almost $150 billion, or just over 2% of world trade. The developed world accounted for two-thirds of this production and consumption.

In many developing countries, forest-based enterprises provide at least one-third of all rural non-farm employment and generate income through the sale of wood products. The value of the trade in non-timber forest products, including pharmaceutical plants, mushrooms, nuts, syrups, and cork, has been estimated at $11 billion.

Source: World Bank (2004)

WORKING THE ECONOMICS OF FORESTRY CARBON PROJECT AND ASSESSING THEIR VIABILITY

Carbon credits provide an extra income from the forestry carbon projects along with other social and environmental co-benefits. Thus, to launch a forestry project and to secure finances, estimates regarding the potential carbon credits from the projects have to be estimated, based on the amount of carbon sequestered by the forestry project. These estimations should be accurate and precise to minimize the risks involved during the developmental processes and in the final outcome of such projects. See Box 6.3.

Many methodologies are available to help calculate accurately the amount of carbon sequestered in forest carbon projects. There are also many easy-to-use models that simulate stocks and fluxes of carbon in trees, soil, and wood products (in the case of a managed forest), as well as financial costs and revenues, and the carbon credits that can be earned under different accounting systems (though mostly suited to American/European countries). These models can give vital details about the estimation of sequestered carbon and other aspects of it. Any project proponents can design their project-specific methodologies, but should have the approval of the CDM Executive Board before using it. Thus, designing one's own methodology can be very expensive and may heavily affect the proposed profit of the project; hence, preparing new methodologies can be feasible only in the case of large or very large projects that can bear such huge costs. For small projects, it is

BOX 6.3 ***Carbon sequestration in forest plantation projects: making CDM attractive to small stakeholders***

Carbon storage in forests is not permanent. The leading paradigm among negotiators of the Clean Development Mechanism (CDM) is that non-permanence should be addressed by an appropriate carbon credit accounting method. A review discusses these methods and presents a model that calculates the minimum area that a hypothetical forest plantation project should be able to cover its transaction costs with the revenues from sales of carbon credits. The model compares different accounting methods under various sets of parameters on project management, transaction costs, and carbon prices. The results of the model show that under the current carbon price and the average transaction costs, projects less than 500 ha are excluded from the CDM, whatever accounting method is used. Temporary crediting appears to be the most favourable approach to account for non-permanent carbon storage in forests and also for the feasibility of smaller projects. However, the possibility of low price for credits with finite lifetime is a risk that could prevent the establishment of CDM forestry projects. Plantation projects with low risk of unexpected carbon loss and sufficient capacity of insuring or buffering the risk of carbon re-emission would benefit from equivalence-adjusted average carbon storage accounting rather than from temporary crediting.

Source: Locatelli and Pedroni (2003)

recommended that various existing methodologies should be followed. (For details see Appendix 7.B)

USING ECONOMICS TO ASSESS FEASIBILITY

Abatement Cost

The first and most critical cost assessed to understand the feasibility of a project is "opportunity cost", also known as abatement cost (cost incurred during abatement of one unit of carbon credits). Abatement costs for the seller are defined as the costs of producing one unit of (uncertified) carbon-sequestration services, or the cost of producing one unit of biomass carbon (Figure 6.1). In any given location, abatement costs can be estimated as the opportunity cost associated with a carbon-

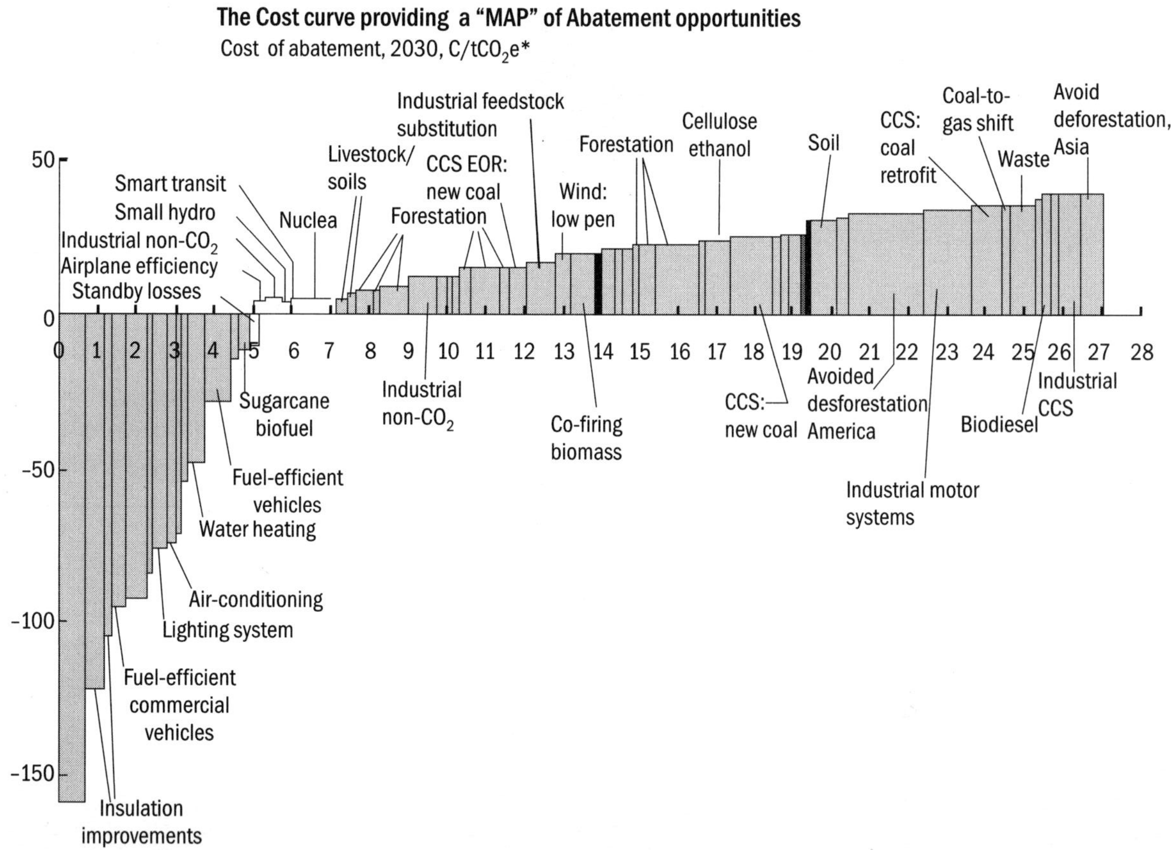

*Cubic feet of carbon equivalents.

FIGURE 6.1 Cost of abatement by different types of projects

Source: McKinsey and types of vattentall analysis

sequestration activity rather than the most profitable alternative activity, or with the switching from the previous land use to the new land use, as represented in Equation 6.1. This cost includes the present value of the stream of revenues and additional risk exposure or loss of food security arising as a result of participation in the project (Cacho and Russell 2006). The opportunity cost, ignoring risk perceptions and other barriers to adoption that can be overcome by participating in the project, can be calculated by the following equation:

$$\mathrm{OC} = R_C,t - R_P,t, \tag{6.1}$$

where R_C,t and R_P,t are the annual net revenues of the current land use and the proposed land use, respectively, and OC is the opportunity cost.

This cost assessment is made essentially at the very early stage of the project to assess the profit to be gained from the project. The abatement cost tells about the estimation of the total incurred cost on sequestering one unit of carbon, which can be used as a reference in calculating the projected benefits of the project. This assessment is also used to depict the profitability of the project to potential investors, with an objective to secure finances. In addition to this initial cost estimation of carbon sequestration (explained in the next section), the execution and management of the project are associated with some other costs, which should also be calculated accurately to have a correct economic feasibility of the project.

FORESTRY CARBON PROJECTS AND ASSOCIATED COSTS

The various elements of cost in a typical forestry carbon project should be assessed properly for the appropriate economic analysis of the project. So far, many models have been developed to predict the cost of carbon sequestration involved in forestry carbon projects. These costs form a very critical component in the calculation of profitability and feasibility, and are hence considered very important. Precise calculation of cost makes a project more viable and feasible. Even the decision of financers and project proponents, which is critical to the health of the project, depends on the accurate estimation of the costs and expenses. To make the estimation easier, the costs of forestry carbon projects are broadly divided into three types: (1) technical costs, (2) costs arising specifically from the carbon component, and (3) costs arising from

BOX 6.4 ***Another way of classifying the cost***

The costs of forest carbon projects can be separated into three areas:

1. The cost of project implementation such as land, initial surveys, ground preparation and planting, early monitoring and treatments, ongoing management and tending, and recurrent forest inventory.
2. The costs of ensuring and increasing the awareness and abilities of local participants.
3. The costs associated directly with the development of the greenhouse gas project documentation, auditing and registration, and then ongoing monitoring, reporting, and verification.

Source: Baalmann and Schlamadinger (2008); FAO (2010)

project management and oversight. These costs have been described in the following subsections (Chenost, Gardette, Julien, *et al.* 2010) while Box 6.4 provides an alternate way of classifying cost.

Technical Costs

Technical costs arise from the operational implementation of a project and vary widely from project to project, depending on their type and objectives. Some capital-intensive projects (such as commercial plantations) may involve very high costs both at the time of plantation and during monitoring phases, and the output value of the plantations justifies the investment. However, other smaller plantations (such as village plantations, fuelwood plantation, or orchards) that generate much less revenue may have smaller implementation costs, which are comparable to the income they eventually generate. Finally, some projects may operate outside the logic of economics and may involve large investment costs and only modest direct income, if any. An example could be "reforestation done to restore degraded lands or to control erosion" (such as soil rehabilitation or erosion control activities), which requires large investments. Although costs vary widely between projects, their structure is fairly similar (Chenost, Gardette, Julien, *et al.* 2010).

Land is an important issue in all forestry carbon projects, and the costs of securing titles to the land make up a large portion of the investment. This huge investment is to be made at the initial phase of the project to acquire or secure lands and to pay for the work required prior to establishing a plantation (production of seedlings, ground and soil preparation, planting, initial and further maintenance). To reduce these initial costs, model contracts can be developed with the participation of landowners, in which the latter make their lands available in return for a fee, either as rent or through participation in the project. Forest plantations can be established with rotations varying from 15 to 60 years in tropical areas, except for some very short rotation species such as Eucalyptus, which can be harvested even after less than 10 years of growth. Success in plantation establishment depends a great deal on how well it is managed in the early years, which explains why investments at the start of the project are so large.

As plantations grow, a phase in which costs decrease, silvicultural operations become less frequent (for example, fewer clearing operations, selective thinning and maintenance activities, and so on). Although maintenance is less intensive, it still requires human resources, material, and equipment (if only for the surveillance of the plantations) deployed over long periods. Therefore, the cumulative costs of this phase should not be neglected as they may eventually (over 20–40 years) amount to more than the initial investment costs.

Transaction Costs

Transaction costs are the costs of participating in the CER/VER market, or converting an emission offset into a CER/VER. In simple words, it signifies the cost associated with the selling of CER/VERs to a party and also the cost incurred while maintaining the carbon stocks during the period of trade and till the course of verification. However, carbon sink forestry projects may be constrained by high transaction cost incurred in developing projects, measuring, certifying, and selling the carbon-sequestration services generated by such projects (Cacho and Russell 2006).

Transaction costs are highly sensitive to the type and size of the projects, especially in the case of forestry projects (Cacho and Russell 2006). The larger the size of the project, the greater the amount of transaction cost incurred by the project. But in some cases where the output of the project is fairly small, transaction cost can be higher on a percentage basis (Chenost, Gardette, Julien, *et al.* 2010). Therefore, to receive certification and enter the CER/VER market, a project will definitely have to internalize various transaction costs. Transaction costs include the following types of costs associated with different aspects of forestry carbon projects (Table 6.1):

Table 6.1 Types of transaction cost

Cost	Buyer	Seller
Search and negotiation	**Ex ante**—find sites, establish contact, organize information sessions, draft contracts, provide training, promotion, establish baseline for region, estimate potential C stocks and flows of project, design individual plans	Attend information sessions, undertake training, design plan
Approval	**Ex ante**—get approval from DNA, validate the project proposal (DOE), submit to CER Board or similar authorities in the case of voluntary project	
Project management	**Ex ante**—buy computer and software, establish office, establish permanent sampling plots **Ex post**—maintain database/software and administer payments, coordinate with field crews, pay salaries, estimate interest costs	Purchase tape and equipment for measuring trees and sampling soil, attend regular project meetings
Monitoring	**Ex post**—enter data from stakeholder sheets, calculate C payments, process soil C samples, obtain verification and certification of carbon by DOE, measure random sample of plots	Measure trees, fill in form and deliver to project office, sample soil C
Enforcement and insurance	**Ex post**—maintain buffer of C, purchase liability, get insurance, settle disputes	Protect plot from poachers and fire

CER – certified emission reduction; DOE – Designated Operational Entity; DNA – Designated National Authority
Source: Adapted from Cacho and Russell (2006)

- Baselines and additionality test costs.
- Risk mitigation (such as insurance, risk management).
- Management costs (administration at the local and regional levels).
- Legal costs.
- Validation and certification of carbon.
- Leakage measurement and coverage (boundary).
- Monitoring of carbon.
- Measurement of sustainable development.
- Consultants for developing projects.
- Hiring of the Designated Operational Entity (DOE) for validation and verification (as the case may be).

Transaction cost thus forms an important part of the overall project cost, and it has been estimated through various case studies that transaction cost constitutes 4%–12% of the overall project cost (Chenost, Gardette, Julien, *et al.* 2010).

Management Cost

Managing, financing, and conducting a project, as well as project communication, also entail significant costs that vary from 8% to 44% of the overall project cost (Chenost, Gardette, Julien, *et al.* 2010). These costs are incurred during the execution and management of the project, which involve meetings, recruitments, and hiring man power to manage the project. It is very difficult to determine a standard figure for the costs of forestry carbon projects, as they depend on the size of the project, costs of labour and land, as well as the availability of skills. See Box 6.4.

Identification of Project Developer and Financial Institutions

We have already seen that the development of a project is associated with many types of costs; therefore, it is very important to identify a project developer who can assist in the development as well as in obtaining funding for the project. Normally, two to five years pass (depending on how fast the project is developed and the payment agreement arrived) before the money is received through the sales of carbon credits. This is because the carbon credits would be paid in such projects only when carbon sequestration is of measurable amount. Therefore, it is important to identify a project developer and donor at a very early stage of the project to arrange for specific early (upfront) payment or compensation arrangements for the involved farmers/planters (FAO 2010).

Table 6.2 gives an overview of the range of costs to be expected during the different steps of a forest carbon project development under different mechanisms (FAO 2010). Box 6.5 gives the computation of the cost REDD.

In a survey conducted by Baalmann and Schlamadinger (2008), also reported by the World Bank, the costs for certification up to the point of registration per afforestation and reforestation CDM project were indicated to the range between $200,000 and $250,000.

Table 6.2 Overview of the range of costs incurred in forestry carbon project

Activity	CDM	VCS	VER+	CCX	Type of cost
Planning phase					
Feasibility study	€15,000–25,000	€15,000–25,000	€15,000–25,000	€15,000–25,000	Consultancy fee
Project documentation	€50,000–100,000	€50,000–100,000	€50,000–100,000	€20,000–40,000	Consultancy fee
Data collection	€5,000–25,000	€5,000–25,000	€5,000–25,000	€5,000–25,000	Internal costs
Validation	€15,000–25,000	€15,000–25,000	€15,000–25,000	Not applicable	Auditor fee
Registration fee	Same as issuance fee, but capped	Not applicable	€550 per year	$5,000 per year	Administrative fee
Initial verification	€20,000–25,000	€20,000–25,000	€20,000–25,000	Not applicable	Auditor fee
Operational phase					
Ongoing monitoring	€3,500–25,000	€3,500–25,000	€3,500–25,000	€3,500–25,000	Internal costs
Ongoing verification	€10,000–20,000	€10,000–20,000	€10,000–20,000	€10,000–15,000	Auditor fee
Issuance fee	$0.10 per tonne of CO_2 equivalent for the first 15,000 per year; $0.20 per tonne of CO_2 equivalent beyond that	€0.05 per tonne of CO_2 equivalent	€150 for up to 1,000 tonne of CO_2 equivalent; €0.03 per tonne of CO_2 equivalent beyond that	$0.12 per tonne of CO_2 equivalent for Non-Annex I countries	Administrative fee

CDM – Clean Development Mechanism; VCS – Voluntary Carbon Standard; VER – voluntary emission reduction; CCX – Chicago Climate Exchange

BOX 6.5 *Computation of the cost of REDD: another aspect of economics*

Figure A illustrates the computation of the costs of Reducing Emissions from Deforestation and forest Degradation (REDD). In this example, avoiding the loss of 1 ha of forest prevents 250 tonnes of carbon from being emitted, but the alternative land use has a stock of 20 tonnes of carbon, so the net emissions avoided are 230 tonnes of carbon.

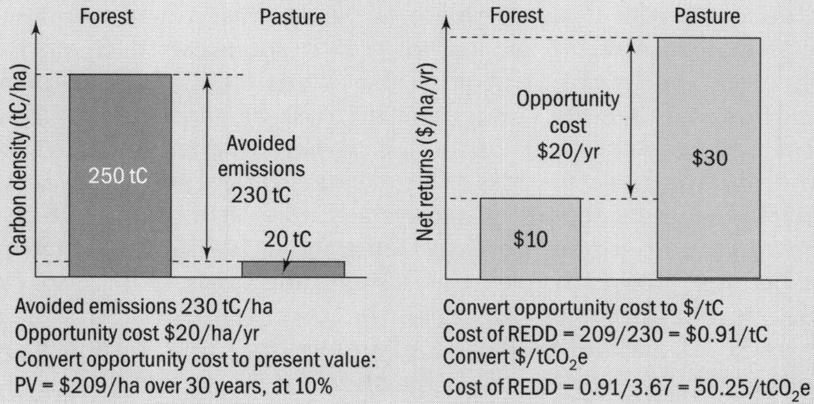

Figure A Computing the costs of REDD

Maintaining forest foregoes $30 per year in income from pasture. Given the $10 per year that forests provide, the opportunity cost is $20 per year. Because this land must be maintained under forest for a long time, this annual opportunity cost must be converted to present value terms. Using a 30-year time horizon and a 10% discount rate, this gives a cost of $209/ha. This is the value of the income foregone by maintaining forest on this land for this time period. To convert this opportunity cost in per hectare terms to one per tonne of carbon terms, we divide it by the net emissions reduction, in this case 230 tonnes of carbon, and obtain $0.91 per tonne of carbon or, equivalently, $0.25 per tonne of CO_2 equivalent. This is the opportunity cost of REDD in this example. To arrive at the full costs of REDD from the country's perspective, implementation and transaction costs must be added. If the implementation costs came to $0.10 per tonne of CO_2 equivalent and the transaction costs to $0.05 per tonne of CO_2 equivalent, then the country would be better off if it were paid any amount over $0.40 per tonne of CO_2 equivalent to

Contd...

Box 6.1 Contd...

avoid deforestation. Individual land users would need to receive a compensation of at least $20 per year to voluntarily forego deforestation.

Note that here, too, what matters is how carbon density changes. Basing the conversion on the entire carbon stock of the forest would assume that it is entirely lost upon deforestation. However, non-forest land uses can often sequester some amount of carbon, so using the entire stock would overestimate the actual reductions in emissions and underestimate the cost of achieving them. In the given example, if the carbon stock sequestered in pasture is ignored, the apparent avoided emissions would be 250 tonnes of carbon per hectare, and the unit cost of achieving it would appear to be $0.84 per tonne of carbon instead of $0.91 per tonne of carbon, or 8% less. The greater the carbon stock in the alternative land use, the greater the underestimation. If forests were replaced with agroforestry with about 100 tonnes of carbon per hectare, for example, the actual net emissions reduction would be 150 tonnes of carbon per hectare, at a unit cost of $1.39 per tonne of carbon, so basing estimates on gross carbon losses would underestimate the cost by almost 40%.

Source: Adapted from Pagiola and Bosquet (2009)

CONCLUSION

Any forestry carbon project can be successful only when it provides intended economic benefits to its stakeholders. Forestry carbon projects thus need to be carefully handled and managed in a way that reduces the overall cost of the projects and enhances the overall benefits (Cacho and Russell 2006). This reduction of cost will increase the profitability of the project and will, therefore, be in a position to deliver the benefits to the communities and proponents involved in the project. Moreover, curtailed expenses will definitely improve the eligibility of the project to get financed. To share the burden of all the expenses incurred in a project, it should be financed by investors.

7

Financial Aspects of Forestry Carbon Projects

H S Gupta

Although investing in forestry projects and activities is an old practice, the possibility of generating carbon credits by prevalent climate change initiatives has dramatically increased investor interest in this sector. Innovative financial solutions are keys to any successful climate change mitigation project in general. Both public and private investors (such as pension funds, insurance companies, and sovereign wealth funds) have the capacity to allocate much of the required capital for such climate change projects, provided that an appropriate risk–reward balance and a proper blending of funding are ensured and the risks are taken by those most able to reduce them.[1] The key challenges in this exercise are as follows:

- How finance is mobilized, of the required scale.
- Innovative financing solutions such as blending and leveraging of private and public sector funds.
- Finance for forestry carbon projects.

Forestry carbon projects provide significant scope for both climate change mitigation and carbon sequestration. Thus, in the near future, forestry carbon projects will be in demand and will play an important role in enhancing the overall effect of mitigation strategies. A secured and effective way of financing is required to develop forestry projects successfully.

[1] Details available at <www.vivideconomics.com>

The kind of projects that are currently being undertaken depends on the potential to use the credits as voluntary offsets or for compliance in mandatory emission reduction regimes, and thus the risks involved can be of great interest to the person who is keen to see the financial aspects.

Some important emerging issues in this fast-growing discipline include the following: role of the public and private sector, bilateral agreements, supply and demand for voluntary/regulated forest carbon credits, favoured standards, price trends for forestry credits, outlook for forestry credits in different compliance and voluntary markets, challenges to overcome, and new approaches.

WHY FOREST CARBON PROJECTS NEED FINANCING?

Forests offer a unique opportunity to mitigate and adapt to climate change. Reducing Emissions from Deforestation and forest Degradation, conserving forest carbon stocks, and enhancing forest carbon stocks through afforestation and reforestation and other measures can help achieve approximately 20% of emission reductions by 2020, which is needed to prevent a rise in global temperature by more than 2°C (Parker, Brown, Pickering, *et al.* 2009).

In addition, tropical forests deliver a globally deployed natural carbon capture and storage service, removing approximately 1 billion tonnes of carbon from the atmosphere annually, for free. Forests also support the livelihoods of 1.4 billion people, directly or indirectly. Thus, forest-owning nations/groups/individuals have better opportunities to adapt to climate change for maintaining the resilience of this ecosystem. But poorer nations, poor communities/individuals, and resource-poor groups will not be able to do this without adequate and predictable financing at scale to move to an alternative low-carbon development path. Hence, for the success of forestry carbon projects, equitable, transparent, and effective distribution of funds will be crucial, taking into account the needs of indigenous and local people. Financing for such forest projects now offers an unparalleled opportunity within the UN climate change negotiations and even beyond that (Parker, Brown, Pickering, *et al.* 2009).

WHAT THE FINANCING SHOULD ATTEMPT AT?

In general, carbon finance is a tool designed to provide a supplementary cash flow to projects (including forestry carbon projects) in which the investment decision is aimed at fostering environmental services and improving livelihoods of the local population, irrespective of the returns from the project. Even some private organizations finance forestry carbon projects either as a measure of self-regulation or due to the imposition of law/regulation, for instance for the regeneration of mining sites or for greening along infrastructure developments. In other cases (both by public and private investors), carbon finance helps improve cash flow of projects not regarded as profitable. Many times, even the additional revenues that projects can realize through selling of certified emission reductions (CERs) or voluntary emission reductions (VERs) do not justify investment, although they somewhat improve the profitability of forestry business endeavours (Neeff and Henders 2007).

Thus, the primary objectives of such specialist financing agencies are to mobilize funds from the developed countries into the developing countries; these developing countries are hosts of the project and they also want to facilitate such projects. These funds try to generate a mixed portfolio of carbon credits, which benefit both the investors and the project proponents. Investors are inclined towards these funds, and they want to procure the carbon credits generated through them. They invest in these funds and obtain the credits proportional to their investment. The project host gets the following benefits (Pearson, Sarah, and Sandra 2006):

- Makes private capital flows available for projects aimed at reducing poverty.
- Facilitates investment in cleaner technologies and best practices.
- Facilitates ongoing partnerships.
- Facilitates support for capacity building for communities and intermediaries.

On the other hand, companies and governments that invest in forestry carbon projects accrue the following benefits:

- Acquisition of emission reduction credits for compliance, trading, and insurance.
- Cheaper transaction costs.
- Risk mitigation via diversification.
- Knowledge of carbon asset creation.
- Demonstration of social responsibility.
- Access to additional credits in each deal.

Additional carbon finance leverages general financing, although it represents a small portion of the total investment.[2] Some key features of carbon finance are as follows:

- With its added revenue stream, carbon finance helps overcome various investment barriers.
- Projects that involve the planting of native species are often not financially viable despite generating carbon revenues; additionality helps in such cases.
- Each dollar of carbon finance catalysed close to $7 of underlying investment. Almost half of this is public funding.
- Domestic sources such as equity from project entities and loans from local banks are major contributors of private investment.
- Although development costs for forest projects used to be higher compared to other sectors, these are gradually decreasing with experience.
- Transaction costs of meeting regulatory requirements can account for about 6% of the total investment for the project; however, this has to be compared with the potential carbon revenues.
- If projects are delayed, the common practice of using carbon revenues to cover operation and maintenance costs could become difficult.
- Carbon revenues can better motivate farmers for maintaining forest use in the long run. See Box 7.1.

BOX 7.1 ***Contribution of finance in increasing land-tenure security in project areas***

With the right institutional instruments in place to clarify carbon ownership and ensure adequate implementation, projects with different land-tenure situations can effectively participate in the Clean Development Mechanism.

Land-tenure securitization can be an incentive for farmers' participation in forestry carbon projects. Finance process can contribute to increasing the level of land-tenure security of forestry carbon projects, but at a definite cost and time.

Source: <www.carbonfinance.org>

---

[2] Details available at <www.carbonfinance.org>

BOX 7.2 ***Financial requirement: the ground reality***

As the contractual payment set-up is often decisive for the economic feasibility of afforestation and reforestation projects, a survey by Merger (2010) asked what kind of financial arrangements project developers require for their projects. Forty four per cent of all project developers indicated that they require upfront financing for their projects. Thirteen per cent said they do not require upfront financing for their projects, while 43% of the respondents did not know or did not provide information about their financial requirements.

These results illustrate that there is currently no standardized method of contracting transactions for forestry credits. This is mainly due to the reason that transactions in the voluntary market are case-by-case and over-the-counter, and that carbon credit exchanges, by contract, are always associated with a project-specific contractual design, which significantly raises the overall transaction costs.

Source: Merger (2010)

TYPICAL FINANCIAL NEEDS OF FORESTRY CARBON PROJECT

To develop a business plan for a forestry carbon project and implement it fully, including marketing needs, a significant amount of investment is required at the very initial stage. This investment provides a broad scope for implementing the project from a business point of view. Financing also provides improved scope of selling temporary CERs and long-term CERs at the prevalent market price. Thus, project proponents have to seek finance at the very initial stage of such a project and they even need support in marketing the potential CERs generated in the course of the project. Box 7.2 illustrates the financial requirements of afforestation and reforestation projects.

Forestry carbon projects are different from energy projects because they require a huge area of land, capital, and labour, and thus initial development is an expensive measure. A vast amount of money is needed to acquire lands for forestry, skilled labour (specific to forestry projects), and other resources such as plants, seeds, fencing wires, and posts.

This large amount of investment and finances are difficult to get at the initial stage of the project. This issue further aggravates if the proponents/ developers of the project are small investors or the project is community initiated. These communities or small investors primarily want to utilize the opportunities available in the forest carbon market along with the restoration of degraded forestland. Thus, these project proponents who evidently do not possess the required capital for a forestry carbon project consequently make it even more difficult to initiate such money-driven projects without the help of external investors.

FINANCIAL CHARACTERISTICS OF FORESTRY CARBON PROJECTS

Some typical financial characteristics of a forestry carbon project are listed as follows (Chapagain and Banjade 2009):

- Forestry carbon projects engage local communities, which may involve relatively higher transactions and implementation costs per unit of emission sequestration compared to other projects; this can decrease the attractiveness of such projects to financers.
- If the project involves a large number of geographically dispersed smallholders with differing priorities and needs instead of a few large-scale landowners or managers, transactions costs may increase, along with the cost of project development or negotiating and enforcing contracts.
- Implementation costs for a project involving smallholders may be higher per unit of emission sequestration. This is because smallholder land uses and management are more varied and often produce fewer carbon benefits than other projects.
- The cost of monitoring carbon benefits may also be higher, and it may be more difficult to credibly establish baseline emissions, that is, business-as-usual emissions, in the absence of the project.
- Therefore, it is imperative that the other aspects and implementation of the forestry carbon project should be made more cost-effective to make it more viable and successful.

Under this backdrop, large-scale forestry carbon projects usually need financing from outside investors and financers, who tend to engage with the project only to make some profit or because of regulations. However, external investors are difficult to be attracted to forestry carbon projects because of the high risks associated with these projects and the complexity of registration and monitoring processes. The temporal nature of the carbon benefits generated by the small forestry carbon projects, coupled with the unstable behaviour of stakeholders (who are, most of the time, native communities attached with land) regarding such projects, creates an impression that forestry carbon projects are unattractive business components. Consequently, well-designed financial plan and investment are required to accomplish the forestry projects successfully (Chapagain and Banjade 2009).

MEETING REQUIREMENTS OF FINANCE AT THE COUNTRY LEVEL

One of the key global challenges in tackling climate change is the need to finance climate change mitigation and adaptation efforts by the developing countries. Current estimates for the total global costs of implementing adaptation and mitigation measures range from $6.8 billion by 2012 to $721.13 million by 2030 (UNFCCC 2010). The Kyoto Protocol and various meetings of the Conference of the Parties (CoP) have articulated this need and have taken measures to address it. Specifically, the United Nations Framework Convention on Climate Change (UNFCCC) has assigned the design and operation of financial mechanisms to the Global Environment Facility (GEF), which is guided by the Subsidiary Body on Implementation in setting its policies, programme priorities, and eligibility criteria for funding. The Special Climate Change Fund (SCCF) and the Least Developed Countries Fund (LDCF) have been established under the UNFCCC, and the Adaptation Fund has been set up under the Kyoto Protocol.

Findings from the recent UNFCCC study suggest that although countries recognize the GEF financial mechanism as a potential financing source, they are hesitant to rely on it. Many countries express concern that financing levels will be variable and subject to voluntary contributions from the developed countries. Despite the acknowledged need for external support to the developing countries in implementing mitigation and adaptation measures, national budgets continue to serve as the primary source of financing at the local and national levels. The burden placed by this on the developing-country economies may prove to be one of the most serious challenges facing the sustainability of climate change measures in the region (UNFCCC 2010). Given the reliance of many countries on national sources of funding, there is a need to further examine mechanisms and sources of internal support. Some suggested potential mechanisms for support include tax revenue, national health insurance levies, and investment from the banking and private sectors.

While domestic financing for climate change exceeds that from external sources, national funding capabilities remain limited. As a result, local-level climate change mitigation and adaptation projects initially intended for national-level scaling up rarely expand beyond project sites, given insufficient budgetary resources (UNFCCC 2010).

There is wide variation in the level of funding being allocated by countries, and it tends to be based on respective climate change contexts, identified vulnerabilities, scope of activities planned, and the status of existing climate change policies, instruments, and mechanisms. At present there are four international financial sources potentially accessible to the developing countries:

1. Multilateral development agencies.
2. Bilateral development agencies.
3. Financial mechanisms of the Convention, including the GEF Trust Fund, the SCCF, the LDCF, and the Adaptation Fund.
4. Overseas development assistance.

One of the main potential sources of funding is the LDCF established under the UNFCCC at the seventh session of the CoP in Marrakesh, and managed by the GEF. The fund addresses the special needs of the 48 least developed countries (LDCs), which are especially vulnerable to the adverse impacts of climate change. Funding available to LDCs includes the development of national adaptation programmes of action, intended to identify urgent and immediate climate change needs.

The UNFCCC has launched a programme to enable financial needs assessments as a part of its National Economic, Environment and Development Study (NEEDS) for climate change projects. NEEDS will provide information to Non-Annex I parties on the implementation costs of mitigation and adaptation measures and will identify appropriate financial and regulatory instruments to support them from the Asia–Pacific region.

Other sources of external funding, identified as existing or potential sources of climate change activity funding, include the United Nations Environment Programme (UNEP); United Nations Development Programme; GEF; funds available under the Convention; Clean Development Mechanism (CDM); Adaptation Fund; SCCF; multilateral agencies such as the World Bank and the Asian Development Bank; regional development funds; the European Community; and bilateral donors such as the Swedish International Development Cooperation Agency and the United States Agency for International Development (UNFCCC 2010).

BARRIERS TO FINANCING FORESTRY CARBON PROJECTS

Some characteristics of forestry carbon projects that pose limitation to investors are as follows (Neeff and Henders 2007):

- **High upfront investment:** Forestry carbon projects involve a vast amount of land and capital, making it economically expensive. This specific characteristic is the long-term horizon of project lifetimes, whereby the bulk of the costs occurs at the beginning of a project mainly due to planting activities and certification in the very initial years (Merger 2010). In addition, project proponents require seed capital, international exposure or technical capacity, credit rating, and access to insurance for developing a project. Thus, it becomes quite a difficult task for investors (particularly the small ones) to invest in forestry carbon projects.
- **Delayed returns on investment:** In contrast to energy projects that deliver returns with the commencement of a project, the high upfront investment in forestry carbon projects delivers returns after a long delay. Carbon forestry projects get the returns when the harvesting of forest commences, and only after that cash flow becomes positive. Revenues become normal only after the forest matures. The lower internal rates of return (IRRs) and various associated physical risks often make such projects unattractive to investors. Thus, in most cases, upfront carbon finance is crucial to overcome such investment barriers (Chenost, Gardette, Julien, *et al.* 2010).
- **Low rates of return:** Forestry carbon projects are designed to provide the dual objective of delivering environmental and socio-economic co-benefits (as CDM projects require). Thus, the direct capital benefits will supposedly remain low, making forestry carbon projects less profitable.
- **High perceived risks:** The success of forestry businesses depends more on the natural conditions around the project boundary such as climate, ecological conditions, and the nearby communities. Thus, these projects are vulnerable to the externalities that are not controlled by human beings.
- **High uncertainties:** The obtained CERs/VERs are of temporal nature, which carry certain uncertainties that may influence projects negatively.
- **Associated complexities:** Additionality is another requirement that needs to be fulfilled before the registration of the forest carbon

project; estimation and calculation of all these aspects is very complex and contentious. So investors remain very cautious while selecting forestry carbon projects for financing.

- **Non-standardized forest carbon markets:** Forest carbon markets are not well standardized in terms of contractual set-up carbon transactions, particularly in the over-the-counter voluntary carbon market where transaction terms vary on a case-by-case basis. Many different carbon payment arrangements exist between project developers and credit buyers (Merger 2010).

ATTRACTING FINANCES

Given the stigma of non-permanence and the leakage of forestry carbon projects (including afforestation and reforestation CDM projects), it sometimes becomes very difficult to secure finances for such projects. Capital- and return-oriented private investors tend to overlook the environmental and social components of forestry projects and thus do not have any genuine interest in supporting these projects. Consequently, public investors are the most active and vibrant financers for forestry projects. Forestry carbon projects are mostly financed through a dual approach of both equity and debt financing. In addition, these projects, after registration, also provide income through carbon credit sales, which provides a "plus" to the IRRs. This income from carbon credits may help overcome the obstacles to investment that are inherent to forestry carbon projects, especially in tropical zones (Chenost, Gardette, Julien, *et al.* 2010).

However, inviting finances is still an art, and this is despite the fact that there are ample opportunities available for financing. In line with this, here are some measures to attract investors into forestry carbon projects; project developers should know about these parameters that are critical for success (adapted from Chenost, Gardette, Julien, *et al.* 2010):

- Adequately assess the needs and time frame of finance.
- Identify reasonable price parameters for carbon credits.
- Develop a selling strategy with an ideal transaction volume and selling price.
- Consider appointing a consultant, lawyer, or market agent who can help develop plans and facilitate complex transactions.

- Establish a methodology to calculate emission reduction and project baselines conforming to evolving rules of the concern mechanism.

The success of a project also depends on the carbon information[3] provided to the financer, which should include the following:

- The full business plan for the project, including operating results, cash flow projection, sensitivity analysis, and risk analysis.
- Financial accounts of the project promoters and main partners.
- A statement of opportunity and/or a feasibility study.
- Permits and/or licenses required to develop the project (ownership of lands, credits, and so on).
- A description of the project and other relevant documents.

WHO ALL ARE INTERESTED IN INVESTING IN FORESTRY CARBON PROJECTS?

With regard to social inclusion and inclusive growth, the co-benefits offered by forestry carbon projects are hugely appreciated in the market. This makes such projects more attractive among various investors (mainly government and commercial companies) who look for inbuilt sustainable development concept in forestry projects. The concept of forestry carbon projects attracts many institutions to invest into such projects by earning corporate social/environmental credits. This also provides ample amount of financing opportunities for the forestry carbon projects, especially under afforestation and reforestation CDM projects. The possible categorization of potential investors is as follows:[4]

- International financial institutions such as the World Bank.
- Environmental non-governmental organizations (NGOs).
- National governments of industrialized countries.
- Regional multilateral development banks.
- Private companies from industrialized countries.
- Philanthropic organizations and international foundations.
- Local commercial banks.

These investors may invest directly in the project or choose a medium to invest. Some of the major media/sources of funding for forestry carbon project development could be in the shape of:

---

[3] Details available at <www.unctad.org>

[4] Details available at <www.forestcarbonportal.com>

- Grants and subsidies.
- Upfront payments for credits.
- Soft loans and loan guarantees.
- Commercial lending.
- Equity investment into companies.
- Investment into other income-generating activities.

Examples of Carbon Financing Institutions

World Bank Carbon Funds are one of the major sources of financing afforestation and reforestation CDM projects. Loan can be procured from these funds, provided it will be repaid when the project is implemented and carbon credits accrued; these funds also facilitate the procurement of CERs thus generated by the project (also see Box 7.3).

These carbon funds assess the scope of the project and its potential capacity to generate CERs/VERs. Most of the time, the financing pattern of a project is driven by the amount of carbon credits it will produce;

BOX 7.3 ***World Bank Carbon Funds***

The World Bank's carbon finance funds help catalyse the emerging global carbon market by extending carbon finance to both developing countries and economies in transition, by linking buyers of carbon credits with climate-friendly projects that are seeking financing. The bank is actively engaged in expanding these opportunities in discussions with other potential Organization for Economic Cooperation and Development carbon buyers—both public and private.

Notes

1. The amount under the Environment Transformation Fund–International Window is non-additional to the World Bank Climate Investment Funds, the Forest Carbon Partnership Facility, and the Congo Basin Forest Fund, as the entire amount donated by the UK fund will be channelled through these three facilities.
2. The World Bank Strategic Climate Fund (SCF) is non-additional to the World Bank Pilot Programme for Climate Resilience, Forest Investment Program, and Scaling-up Renewable Energy Programme for Low Income Countries; these three funds are what comprise the SCF.

finance is provided on a year-wise basis in proportion with the carbon credits generated by the project.

Experience of BioCarbon Fund

The BioCarbon Fund (BioCF), the carbon finance unit of the World Bank, is a public–private initiative that mobilizes resources for pioneering projects aiming at carbon sequestration or conservation in forest and agro-ecosystems, mitigating climate change, and improving rural livelihoods. Figures 7.1 and 7.2 summarize the financial support to afforestation and

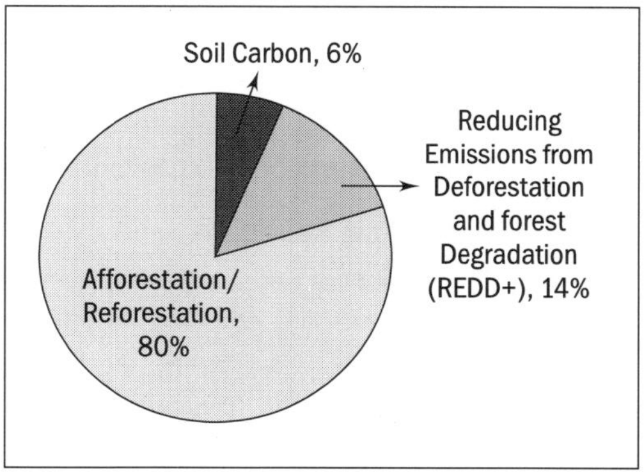

FIGURE 7.1 Financial support to afforestation and reforestation CDM projects by BioCF

Source: Adapted from <www.carbonfinance.org>

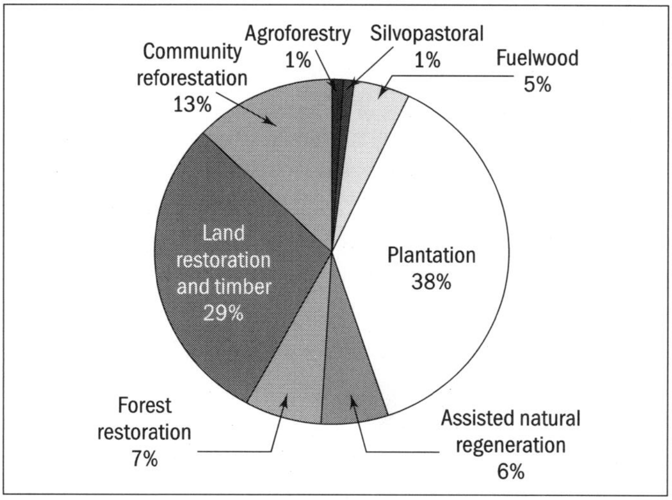

FIGURE 7.2 Percentage of resources committed to different carbon sequestration technologies

Source: Adapted from <www.carbonfinance.org>

reforestation CDM projects until 2011 by BioCF.[5] Most of the BioCF resources are earmarked to afforestation and reforestation projects in different carbon sequestration technologies.

Other Possible Sources of Finance for Forestry Carbon Projects

Several funds and grant programmes (a selective list is given in Table 7.1) exist for carbon sequestration projects. Also some funds or NGOs support small-scale conservation or rural development projects. However, certain indicators of the project idea, such as type, size, target groups, and geographical focus, have to be identified for availing appropriate options (FAO 2010).

Table 7.1 Sources of finance for forestry carbon projects

BioCF of the World Bank	
Description of fund	Based on a public–private partnership model that aims at delivering cost-effective emission reduction and supporting biodiversity conservation and poverty alleviation
Project type	AFOLU projects: afforestation, reforestation, REDD, agriculture
Geographical focus	Open
Case study	• Facilitating reforestation for the Guangxi Watershed Management in Pearl River Basin in China. • Afforestation of 4000 ha, 75% with native species and 25% with eucalyptus. Social benefits are additional employment, direct income increases through sale of non-timber products, and benefits from CER.
Website	<http://wbcarbonfinance.org/Router cfm?Page=Funds&ItemID=24670>
CDCF of the World Bank	
Description of fund	Spreads benefits of carbon finance to the poorest countries and poor communities in all the developing countries, which would otherwise find it difficult to attract carbon finance because of country and financial risk. It is a multi-donor trust fund—a public–private partnership.
Project type	All CDM projects, including AFOLU, are eligible
Target group	Least developing countries—community benefits are a requirement
Geographical focus	Open
Website	<http://wbcarbonfinance.org/Router.cfm?Page=CDCF&ItemID=9709&FID=9709>
German Climate Protection Fund	
Description of fund	The International Climate Protection Initiative has been working since 2008 with annual funds of €120 million. All projects run for 1–5 years.

Contd...

[5] Details available at <www.carbonfinance.org>

Table 7.1 *Contd...*

German Climate Protection Fund	
Project type	• Promoting a climate-friendly economy • Promoting measures for adaptation to the impacts of climate change • Conserving biodiversity with climate relevance (carbon sinks, especially of forests and other ecosystems such as wetlands)
Target group	Projects can be carried out by federal implementing agencies, government organizations, NGOs, business enterprises, universities and research institutes, and international and multinational organizations and institutes, such as development banks, UN bodies and programmes.
Geographical focus	Developing, newly industrializing, and transition countries
Financial support	Between €500,000 and €2,500,000 per project
Case study	• **Mexico:** The project "Climate Change Mitigation in Five Representative Ecosystems" will maintain existing carbon reservoirs in forests and wetlands, and will enable the affected regions and their populations to better adapt to climate change impacts. • **Philippines:** Adaptation to climate change and protection of biodiversity (GTZ, National Department of Environment and Natural Resources)
Website	<http://www.bmuklimaschutzinitiative.de/en/home_i>
GEF Small Grants Programme: Climate Change	
Description of fund	Grants are given towards climate change abatement, prevention of land degradation, and climate change adaptation.
Project type	Removal of barriers to energy efficiency and energy conservation; promoting the adoption of renewable energy by removing barriers and reducing implementation costs; conservation and restoration of arid and semi-arid areas; efficient stoves and biogas to reduce forest loss; integrated watershed management; soil conservation; afforestation; prevention of forest fires; and organic farming
Target group	NGOs and CBOs
Geographical focus	Open
Financial support	Maximum grant amount per project is $50,000, but averages around $20,000. Grants are channelled directly to CBOs and NGOs.
Website	<http://sgp.undp.org/index.cfm?module=projects&page=FocalArea&FocalAreaID=CC>

AFOLU – agriculture, forestry and other land use; BioCF – BioCarbon Fund; CBOs – community-based organizations; CDCF – Community Development Carbon Fund; CDM – Clean Development Mechanism; NGOs –non-governmental organizations; REDD – Reducing Emissions from Deforestation and forest Degradation

The list in Table 7.1 is just indicative and does not cover all the options. Hence, the interested reader may find the following sources also useful for this purpose:

Information on funding sources, policies, and delivery mechanisms, with a particular focus on projects in the developing countries, has been accumulated in the *Sourcebook on Financing for Sustainable Forest Management*. The compilation is based on contents from various sources: donor agencies and countries, Collaborative Partnership on Forests members, international forest-related organizations and instruments, developments banks, private sources, regional processes, foundations, and international NGOs.[6]

Some of the NGOs, foundations, or companies supporting carbon sequestration projects that can be contacted with a well-developed project concept note are The Nature Conservancy (www.nature.org), Flora and Fauna International (www.fauna/flora.org/), Amazonas Sustainable Foundation (www.fasamazonas.org/en/), and the Carbon Pool (www.carbonpool.com).

There are also some other funds that do not necessarily target carbon projects but small-scale conservation or rural development projects. Some investment funds and NGOs are interested in investing in small companies that meet certain conservation criteria, including sustainable land management objectives. A selective list of these funds is given in Table 7.2.

However, the environment for such a specialized finance in the Indian context has not been very encouraging so far, but things are improving very fast (see Boxes 7.4–7.7).

FINANCIAL RISK AND ANALYSIS RELATED TO FORESTRY CARBON PROJECTS

Carbon credits generated by forestry carbon projects are currently being traded at a rate lower than the other energy CDM projects. This is because forestry carbon projects involve high risk due to their peculiar characteristics of being temporary in nature and of ownership. Generally, the risks are borne by both project proponents and financers of the project. This section analyses the risks associated with such projects and their likely impact on the investors (Chenost, Gardette, Julien, *et al.* 2010).

---

[6] Details available at <http://www.fao.org/forestry/cpfsourcebook/en>

Table 7.2 Investment funds interested in small companies

Verde Ventures	
Type of fund	Investment fund of Conservation International
Target	Funds projects that promote biodiversity conservation
Website	http://web.conservation.org/xp/verdeventures/
Eco Enterprises Fund	
Type of fund	Managed by The Nature Conservancy
Target	Works mainly in Latin America and the Caribbean; invests in small and growing, environmentally and socially responsible ventures in sustainable agriculture (including apiculture-, aquaculture-, and community-based energy), sustainable forestry, ecotourism, and non-timber forest projects, as well as carbon, biodiversity offsets, and climate change mitigation and adaptation
Website	www.ecoenterprisesfund.com/index.htm
Root Capital	
Type of fund	Non-profit social investment fund
Target	Pioneering finance for grassroots businesses in the developing world; working with artisan and farmer associations that build sustainable livelihoods and transform rural economies in poor, environmentally vulnerable places
Website	www.rootcapital.org/index.php
Gatsby Charitable Foundation	
Type of fund	Promotes income generation through selected programmes and grants
Target	Supporting basic agriculture and small-scale manufacturing and enterprise in selected African countries
Website	www.gatsby.org.uk/developing.html
Alliance for a Green Revolution in Africa	
Type of fund	Africa-led partnership with initial support from the Rockefeller Foundation and the Bill & Melinda Gates Foundation
Target	Grants for projects and programmes that develop practical solutions to boost farm productivity and incomes for poor. They only give grants for charitable purpose
Website	www.agraalliance.org/

Typical Forest Carbon Project Risks

- Risk associated with carbon instruments such as rejection of methodology or non-compliance to any of the rules mentioned in the Kyoto Protocol/other mechanism.

BOX 7.4 ***Global Environment Facility pledges $250 million for forest finance***

The Global Environment Facility (GEF) has announced a funding envelope of $250 million for sustainable forest management initiatives, including a focus on using forests for climate change mitigation.

Source: <http://env-finance.msgfocus.com/c/15lViFYDBUTXhUnW>

BOX 7.5 ***Relative absence of financial sector: the Indian scenario***

A supportive investment climate for climate change projects requires a well-designed regulatory and policy framework that helps to remove the barriers for implementation and enables efficient utilization of resources by providing positive incentives for efficient options.

One of the concerns expressed by potential investors is that the regulatory framework in India has been uncertain in the recent years, having an adverse impact on the investment climate. Similarly, there has been a reluctance of commercial banks in India to lend for climate-change-related measures, which may improve the bottom line, but may not directly contribute working capital that builds fresh capacity, particularly in the context of renewable energy.

Given this investment climate, very few Indian banks provide lending using a project finance model, which would lend itself well to take into account an additional revenue stream such as carbon finance. More often than not, Indian banks would heavily discount the carbon revenue stream, in part due to the lack of knowledge or due to the uncertainties regarding the carbon market. These banks would not consider emissions reduction purchase agreements as a part of the collateral or guarantee for financing a project with climate-friendly technologies.

Consequently, there are very few indigenous examples to date of the use of financial engineering to improve lending terms, reduce project risks, or monetize or securitize revenue flows from Clean Development Mechanism (CDM) projects. A vast majority of CDM projects are either financed on the balance sheet or financed without taking into account the potential carbon finance revenues, which are seen as an upside.

Source: World Bank (2008)

BOX 7.6 ***Indian initiative for institutional arrangements for carbon financing***

India has proposed to have a new, consolidated fund to be managed by an executive board, operating under the direct supervision and authority of the Conference of the Parties (CoP). The executive board would have an equitable and balanced representation of all parties within a transparent system of governance.

The board would ensure that all parties have direct access to the fund in accordance with the guidelines laid down by the CoP on the policies, programme priorities, and eligibility criteria for accessing the funds. The board may also devolve authority to designated national funding entities of the developing-country parties to approve activities, projects, and programmes for funding, subject to the guidelines and procedures approved by the CoP.

The board would manage a certification and registry system for receiving financial resources from the developed-country parties to be counted towards compliance with financial obligations under the Convention. With the approval of the CoP, the board would institute external independent oversight as well as internal monitoring and evaluation into the management and operation of the fund.

India has also proposed that the financial mechanism would have three separate funding windows aimed at mitigation, adaptation, and technology cooperation. Each of the funding windows would be assisted by a dedicated team of experts in thematic assessment units to carry out the relevant assessments for disbursement to the designated national funding entities of the developing-country parties. The thematic assessment units would be a part of the financial mechanism and would operate under the authority of the executive board.

A trustee selected through open competitive bidding among reputed and pre-qualified institutions would administer the funds. As also stated by the Group of 77 (G77) and China, any funding pledged outside of new financial architecture would not be regarded as the fulfilment of commitments by the developed-country parties under Article 4.3 of the Convention.

Source: Adapted from Parker, Brown, Pickering, *et al.* (2009)

BOX 7.7 ***More innovative funding options***

The Forest Investment Program is a targeted programme under the Strategic Climate Fund to support the Reducing Emissions from Deforestation and forest Degradation (REDD) efforts of the developing countries. It provides an upfront bridge financing for readiness reforms and public–private investments identified through national REDD readiness strategies and other national plans for the management of forests. Additionally, it contributes to multiple benefits such as biodiversity conservation, protection of the rights of indigenous peoples and local communities, poverty reduction, and rural livelihood enhancements.

In addition to the better-known sources of climate-change-related funding, there are a number of examples of innovative in-country initiatives. An example is the Indonesian Low-carbon Development Fund, which aims to leverage and match public and private sources of funding that support low-carbon development.

Source: GEF (2008)

- Risk of non-permanence caused by the fury of nature or human-instigated destructions.
- Risk associated with land rights, lease, community involvement, and sustainable development clause.
- Risk of leakage such as emission from project activities within or outside the project boundary.

Risks in the Phases of the Project

Investors look upon two major things in a carbon forestry project: first, ownership of the carbon credits, and second, technical, financial, and administrative ability of the developers and stakeholders of the project to deliver the promised amount of credits successfully (Chenost, Gardette, Julien, *et al.* 2010). The investors finance a project only when they believe that the project will deliver the intended benefits to all stakeholders. Consequently, they analyse the projects and try to gauge the intensity of the risks associated. Hence, project proponents should make a business plan addressing the concerns of all the investors.

Traditional risks of forestry carbon project can be divided into three phases: planning, construction, and operation risks (UNEP and

Table 7.3 Categorization of risk in forestry carbon project

Planning phase	Construction phase	Operation phase
Feasibility risk	Time overrun risk	Environmental risk
Permit/license risk	Capital cost overrun risk	Technology risk
		Market risk
		Supply risk and sequestration risk
		Operating risk
		Political, legal, and regulatory risks
		Financial risk
		Counter-party risk

Ecosecurities 2007) (see Table 7.3). Generally, after the completion of the planning phase and acquiring of the necessary permits and licenses for a project, a lender will start its in-depth financial assessment. Alternatively, the lender may also discuss about the project viability with a project developer at the initial stage.

- **Planning-phase risk:** These risks are associated with project feasibility, including the potential of the site for relevant forestry project, approval of the local and state governments, and permits or licensing of the site. A detailed survey of the place and knowledge of the characteristics of the geographical area predict the feasibility of a forestry project. Moreover, feasibility study can also be done by comparing the project characteristics with that of the most important criteria for its success/failure. The risk of legal issues can be addressed through an independent assessment of risks by commissioning experts with experience in similar projects. A proper understanding of the relevant regulatory regime is essential and early engagement with the relevant authorities is often desirable (UNEP and Ecosecurities 2007).
- **Construction-phase risk:** The main construction-phase risk is overshooting of the designed schedule, because of which the project might not commission on the required schedule. This indiscipline in project initiation might lead to cost overrun and consequently to low profit margin. These risks can be well managed by having fixed-rate contracts with the construction company and other equipment

providers, in the form of incentives (such as bonuses for timely completion) and/or penalties (such as performance bonds or completion guarantees allowing for monetary damages to be imposed for delays in delivery or completion) (UNEP and Ecosecurities 2007).

- **Operation-phase risk:** Operation phase, as depicted in Table 7.3, comprises many risks related to all types of malfunctions within the project, such as forest fire, technological failure, issues of non-permanence, and so on. The project should be well designed and planned to overcome these types of risks. Some risks in forestry carbon projects such as natural disasters, fire, and political turmoil cannot be dealt with and are thus uncontrollable. Nowadays, insurance is the only solution in these types of situations. In the case of market and counterparty risks, the best option is to have a defined agreement such as the emissions reduction purchase agreement.

How Financers can make Risk Assessment?

As finances are very important to project proponents, so are the returns to financers and investors. Therefore, many tools are available to judge the risks associated with a forestry project. After evaluating the risks of a project and scrutinizing the assessment report, the financers decide whether to invest or not. Following are the steps taken to assess the risks associated with a particular forestry project (adapted from UNEP and Ecosecurities 2007):

- **Risk identification:** In this step, all the associated risks are identified. Either project sponsors or independent experts carry out the identification. However, lenders usually commission an expert risk analyst (such as an insurance company involved in the project) to undertake this activity.
- **Risk matrix:** A matrix is drawn to plot each risk against the phase of the project in which it occurs, its likely scale of impact, and the parties affected by the risk, and with its mitigating procedure. This generally forms the basis of negotiations between parties as to the apportionment of the various risks (UNEP and Ecosecurities 2007).
- **Quantitative risk assessment:** After the above two steps, a quantitative risk assessment should be carried out on the project as a whole. A quantitative estimate of the total value at risk or a

comparative risk index (compares the risk of a project with that of other similar projects) may be obtained as the output.

This assessment may then be modified to discount the absolute risk[7] by a factor reflecting the availability of risk management options to reduce either the likelihood of an event occurring or its impact.

What Financers Should do to Mitigate the Risks?

There are essentially three options for managing risks:

- **Change the project:** If a risk has been identified and understood in the early planning stages and found to be unavoidable, then the only possible solution is to change the project to minimize the risk.
- **Allocate the risk to the most appropriate party:** This solution talks about distributing the risks usually to the entity that is most closely associated with a risk such that it can bear that risk at the lowest cost. An example of such a solution is giving the contract of plantation to a party that can ensure a successful plantation.
- **Transfer the risk to a third party:** Financial instruments such as hedging, third-party guarantees, or insurance may be used to transfer risks to third parties. Hedging involves the use of derivative markets, for example, to fix the future prices of commodities, currencies, or interest rates. Export credit agencies or international institutions such as the World Bank's Multilateral Investment Guarantee Agency have the authority to provide third-party guarantees. Insurance is just a premium-based mechanism to help the insurer bear the risk of uncertainties related to the project (UNEP and Ecosecurities 2007).

CONCLUSION

Thus, to provide for the expenses incurred in different activities/phases of the forestry carbon project, the project promoter must have an accurate idea of the finances involved and of how they can be sourced. In addition, if the project proponents are well acquainted with procedures and sources of finances available, they can provide the necessary information

---

7 Absolute risk is a measure of the risk posed by a specific event without countermeasures being taken. It is defined as the product of two factors: the likelihood of an event occurring and the significance of its impact (if it does occur). Past records and professional judgements may be used to provide quantitative data for both factors. "Significance" may either be an index (such as, a scale from 1 to 10) or a monetary amount (damages) (UNEP and Ecosecurities 2007).

generally sought by the investors to finance a project. Consequently, the choices made by investors will largely depend on the confidence they will have in the smooth running of the project. This is also assessed through an analysis of the "risks" associated with the project.

Financial risk is inherent to a forestry carbon projects, which is bound to be risky. This is despite the fact that a forestry carbon project can provide good returns because of its socially inclusive nature and co-benefits to the communities involved. Thus, a forestry project needs to be designed with more care and more insight, particularly in the planning phase of the project. Adherence to laws and regulations can improve the probability of the success of the project. Good project proponents will thus make sure that the project meets all the required criteria of being ideal and thus reduce the risk exposure of all the stakeholders, including financers. An accurate assessment of the risks associated with finance, along with a well-designed procedure, can attract the potential investors into the forestry carbon project and can thus play a vital role in the success of the concerned project.

8

MARKETING ASPECTS OF FORESTRY CARBON PROJECTS

M Yadav and H S Gupta

SCIENCE OF CLIMATE CHANGE AND BUSINESS OF CARBON

While the world scientific community under the aegis of the Intergovernmental Panel on Climate Change is still struggling to put forth convincingly its scientific evidences in support of changes in the climate scenario and its implications for earth and on humanity, the business community is by and large focused on economic interests. The Copenhagen Accord in 2009 under the aegis of the United Nations Framework Convention on Climate Change (UNFCCC) could ensure a political commitment of the world leaders to reduce greenhouse gas (GHG) emissions by 50% of its level in 1990 and to restrict global temperature rise to a maximum of 2°C only by 2050. However, there is no concrete commitment emerging on the Kyoto Protocol beyond 2012 when its first commitment of 2008–12 came to an end. The hope for any binding agreement post 2012 has not materialized even at the Cancun and Durban conferences of 2010 and 2011.

While the world political leadership may be dithering on immediate commitments on GHG reduction forced by domestic compulsions, and the part of the scientific community may give different opinions that might help provide justification for political positions, the fact remains, as believed by the larger world community, that the anthropogenic climate crisis is real and needs emergency solutions. This realization has led to business organizations commit them in addressing climate crisis by strategies that not only reduce GHG emissions but also make them zero carbon. This trend has contributed to the growth of both regulatory and voluntary carbon markets. Experts feel that this brings a lot of potential to the future of carbon business from general and forestry sectors in particular, although countries have not agreed so far on a binding agreement for the foreseeable future.

This chapter will discuss the broad aspects of carbon markets and their types, the supply chain, and the marketing aspects of all types of forestry projects, including afforestation and reforestation Clean Development Mechanism (CDM) and voluntary forestry carbon projects.

UNDERSTANDING THE CARBON MARKETS

Carbon market refers to the buying and selling of emission permits that have been either distributed by a regulatory agency or generated by GHG reduction commercial projects from any one or a combination of six GHGs (carbon dioxide, methane, nitrous oxide, sulphur hexafluoride, hydrofluorocarbons, and perfluorocarbons). An emission reduction of GHG equal to 1 tonne of CO_2 equivalent (tCO2e) is called one carbon credit and is expressed as 1 tCO_2e; this is the trading unit in carbon markets. GHG reduction credits can be accrued through two different approaches/transactions: (1) project-based transactions, in which emissions credits are obtained from a specific carbon offset project; and (2) allowance-based transactions, which involve the trading of issued allowances created and allocated by regulatory agencies under a cap-and-trade regime. Under the latter category, the regulatory authority puts a cap on the permissible quantity of emissions by the participants (country or organization) and issues a number of assigned amount units equal to the cap. In the cap-and-trade carbon regime, the entities emitting less than the permitted units can sell the balance to those emitting more than the permitted units. In another way, carbon markets can be classified into two major categories:

1. **Compliance or regulatory markets:** The carbon credits generated through the compliance market under the Kyoto Protocol's CDM are known as certified emission reductions (CERs).
2. **Voluntary markets:** The carbon reductions that are not regulated or not sold through a compliance mechanism (Kyoto Protocol or any other national compliance mechanism) are known as verified emission reductions (VERs), also called voluntary emissions reductions (see Table 8.1 for details). Since the voluntary market inherently does not

Table 8.1 Difference between CERs and VERs

CERs	VERs
Compliance markets generate and trade GHG emission reductions known as CERs that are regulated and directly initiated under the Kyoto Protocol's CDM.	Voluntary markets generate and trade GHG emission reductions that are not regulated or directly initiated by the Kyoto Protocol and are known as VERs.
GHG reductions of any CDM project are measured according to internationally agreed methods and are quantified in standard units called CERs. These are expressed in tonnes of carbon dioxide (CO_2) equivalents.	GHG reductions outside the Kyoto Protocol are measured according to the internationally agreed methods and are quantified in standard units called VERs. These are also expressed in tonnes of carbon dioxide (CO_2) equivalents.

CER – certified emission reduction; CDM – Clean Development Mechanism; GHG – greenhouse gas; VER – verified or voluntary emission reduction

operate under a cap-and-trade system (where emission allowances can be traded), all can be referred to as the "voluntary offsets market" and the carbon credits so generated or earned are referred to as carbon offsets. Further, voluntary carbon transactions outside the Chicago Climate Exchange (CCX) do not operate via a formal exchange; it can also be referred to as voluntary over-the-counter (OTC) market.

A forestry project is one through which carbon has to be captured from the atmosphere in the form of carbon sequestration by trees. The CDM under the Kyoto Protocol permits a maximum of 1% CDM credits from land use, land-use change, and forestry (LULUCF) methodologies. In addition to these Kyoto markets, the forestry projects have voluntary OTC markets as the only other outlet. Although LULUCF projects face the problem of lower financing, and technical and bureaucratic hurdles under the CDM, they are relatively popular options in the voluntary OTC markets because of their inherent values and benefits to the local communities and the biodiversity, and other values that voluntary buyers appreciate (because many ultimate consumers have an understanding of the role trees play in the carbon cycle and of other associated benefits; and this sentiment guides the market).

Market-based incentives and regulatory compliance, as well as a combination of the two, are both necessary and thus have been advocated as measures to mitigate the climate change impacts through GHG reductions. Markets for emissions reductions though forestry projects are among the most innovative and cost-effective means of creating a market pull for carbon reduction, while, at the same time, providing incentives to people in varied manner.

The compliance markets can be further classified into international compliance, regional compliance, national compliance, and local/state/ province-level compliance markets. These markets are discussed in the following subsections. Figure 8.1 summarizes the diversity of carbon markets.

Compliance Carbon Markets of International Level

Only one global compliance treaty, the Kyoto Protocol, is operational at the international scale that sets binding targets for the Annex I (developed) countries to reduce their GHG emissions. Thus, only one international compliance market is associated with this treaty. The bases for the regulated international compliance carbon market are three flexible mechanisms created under the Kyoto Protocol to provide the Annex I countries with a cost-effective means of achieving their GHG emission reduction targets. These bases are emissions trading (cap and trade), joint implementation, and CDM. However, the CDM under the

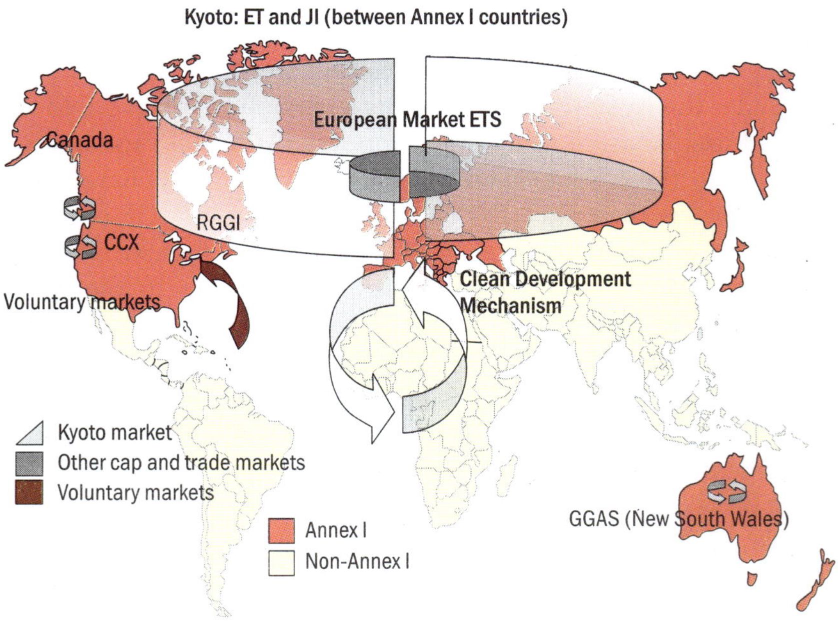

FIGURE 8.1 Carbon market diversity

Source: CDM Executive Board under UNFCCC

Kyoto Protocol permits forestry carbon projects under the afforestation and reforestation methodologies limited to 1% of total CDM credits.

Compliance Carbon Markets of Regional Level

The European Union is the only regional group that also has regulation for GHG reduction commitments for its member countries, and there exists one compliance market for this group of countries. The GHG reduction targets for this regional group are set above the Kyoto Protocol targets; thus, the Kyoto Protocol targets are automatically achieved when the regional targets under the European Commission are achieved. The largest of these is the European Union Emission Trading System (EUETS) set up in January 2005 to help the European Union region achieve its Kyoto Protocol targets. However, the EUETS does not allow forestry projects (Bettinger, Clutter, Siry, *et al.* 2009).

Compliance Carbon Markets of National Level

Although some countries have not ratified the Kyoto Protocol and thus have not taken the legally binding GHG reduction targets under the international compliance regime, they have taken initiatives at the national level. The USA is the only Annex I country that has not ratified the Kyoto Protocol and thus does not have legal GHG reduction commitments so far.

In 2006, nine states and territories of Australia had proposed a National Emissions Trading Scheme to reduce emissions by 60% by 2050. The scheme also allows forestry projects located within the country. However, in 2007, the Government of Australia established the International Forest Carbon Initiative, with the pledge to help reduce deforestation in the Asia–Pacific region by investing $160 million in association with other countries that are willing to contribute to this fund.

The Government of Norway has set up the Norway Government Fund to purchase emission credits based on the flexible mechanisms of the Kyoto Protocol, so that the country can reduce its GHG emissions by 40% by 2020 below the 1990 level and then become carbon neutral within another 10 years, that is, by 2030. The Norway MoF plans to buy approximately 27 million tonnes for delivery from both forestry and joint implementation projects during 2008–12.

The UK initiated a pilot trading scheme, called the UK emissions trading scheme (ETS), in 2002 prior to the mandatory EUETS coming into existence, but the scheme was closed in 2006.

Local/state/province-level Compliance Carbon Markets

Although the USA, the largest per capita emitter of GHGs in the world, has chosen not to ratify the Kyoto Protocol to date, many of its states have taken initiatives to regulate emissions. The state of Oregon enacted the Oregon Standard for new power plants to reduce their CO_2 emissions by 17% below the most efficient combined cycle plant, the first regulation of CO_2 in the USA. See Boxes 8.1 and 8.2.

On the East Coast of the USA, eight states developed the Regional Greenhouse Gas Initiative (RGGI) to commit participating states to

BOX 8.1 ***India's own domestic initiatives***

India has envisaged setting up of a national trading mechanism for energy efficiency certificates under the Perform, Achieve, and Trade (PAT) scheme in which companies using a large amount of energy can reduce their energy consumption and earn energy certificates. Moreover, if they achieve more than the set level of efficiency, they can sell the additional certificates to those who achieve below the set level.

The Bureau of Energy Efficiency under the Ministry of Power, the agency that initiated this scheme, will launch two exchanges to help companies register and trade certificates. The scheme will be similar to the Clean Development Mechanism and even companies that reduce their power consumption can register with the United Nations. India targets a reduction of 10 million equivalents of oil consumption in the next three years, and it has identified nine energy-intensive sectors to target in the first phase under the National Mission on Enhanced Energy Efficiency. However, adopting PAT is voluntary, so far in nature.

The Ministry of Environment and Forests launched a second scheme, the Emissions Trading Scheme (ETS), in September 2011 where in companies in high emissions sectors will be given targets to cut their emissions and trade through ETS the surplus from their targets. The Central Pollution Control Board will be the implementing agency for the scheme.

Source: *Hindustan Times* (27 January 2011)

BOX 8.2 ***Chicago Climate Exchange***

The Chicago Climate Exchange (CCX) is the world's first voluntary, legally binding greenhouse gas reduction and trading system. The CCX relies on independent verification and has been active in trading emission reductions since 2003. Exchange members who cannot meet their own emission targets can purchase credits from other members who exceed their emission reductions, or have verifiable offset projects. Forestry projects are eligible under CCX rules, and the exchange accepts forestry emission reductions originating from forests established after 1990. These forests have to be managed sustainably, and third-party forest certification is required. Individual projects are audited and reports prepared annually. Recognized carbon pools include above- and below-ground living tree biomass and long-lived wood products. The CCX requires that 20% of credits registered annually be kept in reserve to cover potential shortages at the conclusion of a forestry project.

Source: Bettinger, Clutter, Siry, *et al.* (2009)

cap their GHG emissions at the 1990 level by 2009 and then reduce them by 10% until 2018. The RGGI covers electric utilities larger than 25 MW, which can use offset projects to meet reductions but on a limited basis. RGGI allowances for the period 2009–11 were sold for \$1.86, as announced by the RGGI Inc. on 10 September 2010.[1]

On similar lines, the state of California passed a bill called AB 32 requiring 25% cut in CO_2 emissions by 2020. However, the bill is under reconsideration owing to the financial crisis that erupted in the USA in August 2008 and the resultant unemployment across the country. The state also developed the California Climate Action Registry (CCAR). The Climate Action Reserve (CAR), the parent organization of CCAR, tracks and registers voluntary projects that reduce emissions of GHGs. Forestry offsets are permitted under the CCAR programme.

California's Air Resources Board (ARB) gave the official go ahead to the state's cap-and-trade plan on 16 December 2010. The decision is the culmination of an eventful year for California's AB 32 legislation, which aims at reducing the state's GHG emissions to 1990 levels by 2020. The offsets will come from the 2005–14 vintages of CAR credits from

---

[1] Details available at <www.pointcarbon.com>

projects related to methane digestion, destruction of ozone-depleting substances, urban forestry, and forestry (Carbon Positive, December 2010).

The state of New South Wales (NSW) in Australia started the NSW Greenhouse Gas Abatement Scheme (GGAS) in 2003, which required energy producers to restrict their emissions within 8.65 tonnes of CO_2 equivalent per capita in 2003 and to further reduce this limit by 3% to 7.27 tonnes by 2007. The energy producers can offset their reduction targets by purchasing NSW Greenhouse Abatement Certificates or by paying a fine of $11 per tonne. The GGAS penalty for benchmark participants that fail to reduce emissions has risen from $1.50 to $14.00 per tonne of CO_2 equivalent, effective for the 2010 compliance year.

In southern USA, the state of Georgia established the Georgia Carbon Sequestration Registry, the first carbon registry in the region. The purpose of the registry is to encourage actions that reduce GHG emissions and to ensure that Georgia forestland owners receive proper consideration for their management activities in the emerging carbon markets. The ownership of carbon is not permanently tied to the land or trees, since sequestered carbon is treated as a separate commodity that can be traded independently of wood. Registered forests must be located in the state of Georgia, composed of native tree species, and managed according to Georgia's Forestry Best Management Practices (Bettinger, Clutter, Siry, *et al.* 2009).

The Climate Trust, a non-profit organization established in Oregon in 1997, is dedicated to providing solutions to problems of climate change. This trust invests in high-quality carbon offset projects, including those that are related to forest management activities. It also promotes the development of carbon policies on local, regional, national, and international levels (Bettinger, Clutter, Siry, *et al.* 2009).

In addition, the Oregon House Bill 2200 authorized the state forester to establish programmes to market forestry carbon projects. A proposed carbon accounting system (registry) will be administered by the Forest Resource Trust (Bettinger, Clutter, Siry, *et al.* 2009).

Voluntary Carbon Markets

Carbon transactions that are not required to meet the legally binding commitments but are undertaken for marketing or social responsibility purpose are known as voluntary carbon markets. Such transactions can take place either through an exchange such as the CCX or without a formal exchange. The second type of transaction is called OTC voluntary carbon markets. In fact, voluntary carbon markets predate all regulated carbon markets. The world's first carbon offset deal was made in 1989 (well before the signing of the Kyoto Protocol in 1997 and long before it came into effect in 2005), when an agroforestry project in Guatemala was sponsored by a US electricity company AES Corp. (Hawn 2005). AES wanted to reduce its "carbon footprint" for philanthropic and marketing reasons, through offsetting the GHGs it emitted during electricity production by paying farmers in Guatemala to plant 50 million pine and eucalyptus trees on their land. This deal made the beginning of a voluntary carbon market that still remains as debatable and interesting as it was in 1989. Since voluntary carbon markets are negotiated on a case-to-case basis, and many of the deals do not require a uniform certification or verification process for earning carbon credits or registration with any central body, these are subject to criticism from environmentalists as merely "a game of smoke and mirrors" rather than an engine of actual emission reduction agenda. Despite the shortcomings of the voluntary carbon market, these do not incur the type of transaction costs the CDM projects have. For example, the cost of getting a carbon offset project approved by the CDM Executive Board ranges anything between $50,000 and $250,000 (Krolik 2006). As calculated by the United Nations Development Programme, by the time a typical small-scale CDM project gets registered creating CERs, the total upfront costs will account for 14%–22% of the net present value of its revenue from the carbon credits (Krolik 2006). Compared to this, voluntary carbon markets involve no or low transaction costs and provide flexibility to both buyers and suppliers, particularly small firms and individuals who contribute 50% of GHG emissions.

MARKETING OF FORESTRY CARBON PROJECTS FOR CERS/VERS

The compliance market specifically related to forestry projects are designated as afforestation and reforestation CDM projects under the Kyoto Protocol. A compliance market such as CDM invites almost all the afforestation and reforestation projects that comply with the legal obligations and procedures defined under afforestation and reforestation CDM. On the other hand, voluntary market, as discussed earlier, is scattered; therefore, the projects should be designed according to the market it intends to target.

Marketing means selling a designed product or creating a market for products obtained in the process. In the context of forestry projects, marketing is required to explain the various features such as sustainable development, additionality, and longevity of the project, which are very well sought and valued by the buyers and the concerned market; thus, marketing is relevant in both voluntary and compliance markets.

The carbon market at present seems to be strewn with complex procedures and imbalance as far as certified emission reduction (CER) trading is concerned. The parties involved or the project developers engaged in trading are still oblivious to an adequate approach towards securing a safe transaction. For instance, the sway of CDM market is still controlled by buyers who consist of mainly governments and banks, primarily because of the risks involved in transaction being very high. In addition to abundant supply of CERs, there is no institutional back-up source for sellers such as the World Bank and for buyers (Cornland 2006).

The presence of such complexities makes project proponents strive hard in fetching the best price for their project and thereby securing higher profitability and project viability. Hitherto, private companies have refrained from indulging much into afforestation and reforestation CDM activities due to the prevailing uncertainty as a result of the long life of forestry projects and their economic viability. Also currently, most of the buyers of CERs are governments, banks, and representative of national governments; therefore, it is a prerequisite to know the expected time of the transaction, preferences of the buyer, speculation of the market, and opportunity cost. Knowing the nuances of the CDM market and its marketing implications, coupled with adequate marketing strategies since the very inception of the project, can help project developers overcome all these problems and anomalies and further assist them to make some critical decision reflecting into increased profitability of the project.

Supply Chain of Forest Carbon Credits/Product

A typical forestry carbon project supply chain comprises four stages: project or project idea generation, GHG reduction verification/certification/approval, selling CERs to distributors, and CER consumption by compliance/voluntary buyers. These steps are shown in Figure 8.2 and explained in the following sub-subsections.

Stage I: Carbon credit product creation/design

Theoretically, a single forest owner might develop an afforestation and reforestation project on his or her land, get it vetted by the Designated National Authority (DNA) and the approved validators, get it approved from the CDM Executive Board, and sell the resulting CERs directly to the buyer. However, in practice, the project developers are generally buyers from the Annex I countries, consultants, private profit or non-profit organizations, and/or governments of Annex II countries, each looking for specific co-benefits (for example, benefits for biodiversity, for the community, or for the industry/individuals) in addition to the basic objective of GHG reductions through carbon sequestration. As discussed earlier, CDM projects are placed under "small" and "large" project category. The "large" projects have dominated the carbon markets as these projects often reduce transaction costs per credit through "economies of scale", thus causing more GHG reduction and generating higher profits at lower prices.

On the basis of durability, the CDM afforestation and reforestation projects are further classified under two major categories: temporary CERs (tCERs) and long-term CERs (lCERs). From the marketing point of view, projects that deliver tCERs receive carbon credits in a 5-year interval after verification. Therefore, they can take advantage of price fluctuations over time by commercialization of issued tCERs, which give the flexibility to sell these credits whenever prices are high.

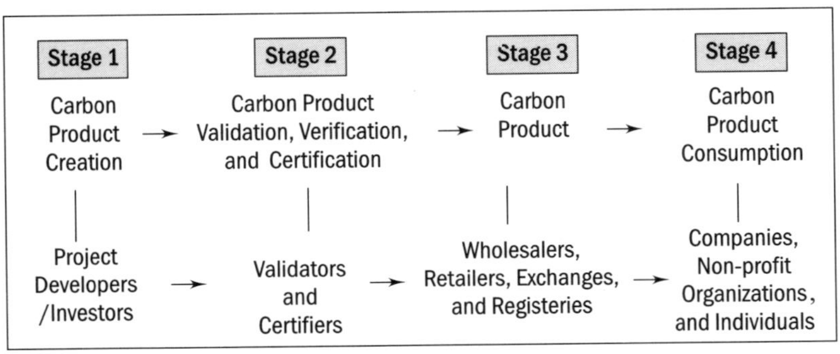

FIGURE 8.2 Simple diagram of a retail carbon market supply chain

Source: Adapted from Bayon and Hamilton (2007)

However, in practice, tCERs are sold only once at an initial stage of the project. Hence, the obtained price can be either advantageous or disadvantageous, depending on future market developments. Thus, project proponents may lose revenue if the prices increase after the deal of tCERs and gain if they decrease in the future (Neeff and Henders 2007).

Although temporary crediting (tCERs and lCERs) is an approach to address non-permanence of afforestation and reforestation projects, it has some market- and management-related implications, as follows:[2]

- Many projects opt for tCERs because the long life of lCERs makes it difficult for developers to take advantage of carbon price speculation.
- The price of forest credits is determined by the need to replace them with a permanent credit, which may be different from the real costs of some projects.
- A replacement rule such as the EUETS leads to non-fungible credits, thereby affecting negatively the demand for forest carbon credits in the market.
- As temporary credits cannot be renewed beyond a project crediting period, the long-term carbon sequestration goals of some projects get limited.

Stage II: Carbon credit product validation, verification, and certification

The second stage in the supply chain begins with creating product credibility; the market recognizes the product through a process of legitimate third-party validation/verification. In the case of a voluntary project, even self-verification can be acceptable to some buyers, given the credibility of the product developer. However, in the CER market, only validation, verification, and certification by CDM Executive Board-approved entities called Designated Operational Entities (DOEs) are acceptable. Some DOEs operating in India are Det Norske Veritas (DNV), Société Générale de Surveillance, and International Maritime Organization. Quantification and verification of GHG emissions reduction require significant technical expertise and monitoring throughout the project life span in terms of accounting of GHG reduction period, payback time, and the amount of GHG destroyed, displaced, or stored.

[2] Details available at <www.carbonfinance.org>

A number of methodologies are used to compute the baseline and the annual increments (which then gets converted into carbon credits minus leakages). Further, a number of issues such as additionality, permanence, leakage, ex-ante and ex-post accounting, double accounting, and co-benefits guide the considerations for offset quality. Generally, project developers bear the cost of verification. The process of verification and certification is much more rigorous in CER projects than in verified or voluntary emission reduction (VER) projects, and hence the transaction cost is significantly high for the former (14%–22% of the net present value of benefits for small CER projects). Also the time taken for project verification by the host country (DNA) and certification (approval) by the CDM Executive Board is quite longer. The inherent complexities of afforestation and reforestation projects such as high transaction costs, time for verification and approval, and the associated risks are all typically higher than any other offset projects; that is why, afforestation and reforestation projects are less preferred by both the project developers and the investors. (The process of verification and certification for CER projects is developed, operated, and controlled by the CDM Executive Board as per the policy framework decided by the Conference of the Parties under the UNFCCC.)

Because of high transaction costs and quality issues arising out of a wide range of product offerings in the voluntary markets for the forestry projects, several certification standards and programmes have emerged to assure buyers about the project quality. But they are also causing confusion among the buyers because of their proliferation (see Appendix 5 for details). However, certification could be a beneficial tool for ensuring consistency in quality, reducing risks for the buyers, and building consumer trusts.

Stage III: Carbon credit product distribution

Once carbon credits are verified and/or certified (approved by the CDM Executive Board in the case of CERs), intermediaries step in either as buyers interested in purchasing credits for resale or as facilitators arranging transactions between buyers and sellers for a service fee. In the case of CERs, if the project initiator itself is the buyer, then it uses the issued credits to meet its reduction targets in the home country. Various carbon market intermediaries include retailers, brokers, investors, exchanges, and registries. See Boxes 8.3 and 8.4.

Wholesalers/retailers

Wholesalers/retailers are mostly profit-oriented organizations that purchase and keep an inventory of carbon credits to resell these to buyers for profit. A retailer's project portfolio can change over time as it purchases both emission reductions and sequestration projects, that is, CERs, VERs, and gold standards. A few such organizations include Carbon Planet and Carbon Neutral to name a few. Many retailers provide carbon credits online to individuals and/or small businesses to become "carbon neutral". These web-based retailers (such as Carbon Neutral) help individuals and/or small businesses compute their carbon footprints using carbon calculators and then offer means to become carbon neutral. It is very convenient for individuals or organizations with low carbon intensity to compute their carbon footprint and purchase the required amount of carbon credits to neutralize their emissions, and thus solving the problem partly.

Investors

While large institutional investors fund the regulated carbon markets, some carbon funds also invest in the voluntary markets, such as the London-based Cheyne Climate Hedge Fund. It is world's first voluntary carbon offset fund. RNK capital and Climate Change Capital are some other offset funds. For details see Chapter 7.

Brokers

Brokers are the individuals or small firms involved in facilitating transactions between institutions and offset project developers, mostly matching buyers and sellers for CERs purchased under the CDM.

BOX 8.3 ***Exchange for carbon credits***

The Chicago Climate Exchange was the first such exchange that started trading of voluntary carbon credits in 2003. Its access is limited to members only. A large number of other exchanges have come up in the last few years such as the European Union Emissions Trading Scheme, which is the largest exchange today. The Bank of New York has also created "centralized, secure, paperless" carbon credit trading services for corporate buyers and sellers. The commodity exchanges across the world have started carbon trading at their exchanges. The Multi Commodity Exchange and the National Commodities and Derivatives Exchange of India have also started carbon trading in India.

BOX 8.4 ***Basic requirements of projects for trading in CCX***

All projects must meet the basic market requirements before trading carbon financial instruments on the Chicago Climate Exchange (CCX) market to qualify as an offset project. The basic requirements that a forestry project proponent must keep in mind are as follows:

- The project owner must have sufficient ownership control to have influence over carbon stocks and sequestration levels for the future.
- The project must be capable of sequestering levels of carbon and generating gross revenues in excess of market start up and participation costs.
- The project owner must be willing to commit to maintaining carbon stocks for the commitment period associated with the CCX market.
- The project must have been certified as sustainable by an independent third party under a programme endorsed by the Programme for the Endorsement of Forest Certification, such as the Sustainable Forestry Initiative, Forest Stewardship Council, or Group Tree Farm System.
- The project owner must be able to track and report on carbon stocks annually in conformance with market rules. This includes the use of a CCX-approved growth model.
- The project owner must disclose any significant emission sources under his or her influence, ownership, or control.
- Carbon stocks and net sequestration must be disclosed on all interests associated with the project owner. For instance, all lands owned must be accounted for and disclosed to the market.
- Project baselines must be developed using a site-specific inventory with statistical standards of ±10% at 95% confidence for the total tree volume. Gaps in the accounting for the volume of all stem of 1 inch diameter at breast height (DBH) and greater may be addressed using other means, with prior approval. (While this is not a requirement under the current CCX market rules, Forecon EcoMarket Solutions has taken the position that these statistical standards are the benchmark from which one can establish project-specific accounting to acceptable tolerances.)

Contd...

Box 8.4 *Contd...*

It should be mentioned here that not all of these mandates must be met to submit a project proposal for review and approval by the market. Some requirements can and must be fulfilled after project approval but before credit registration and trading. Project owners must understand and be willing to assume the risks associated with entering the carbon market.

However, they also provide trading services for VERs in the voluntary carbon markets. They charge a commission of roughly 7.5% for their services. Natsource, CO_2e Capital, Ecosecurities, Trexler Climate + Energy Services are some examples of brokers.

Exchanges and registries

A "commodity exchange" facilitates sale/purchase of a given commodity in required quantity and quality at mutually agreed price, with transparency and guarantee of delivery. Thus, commodity exchanges provide freedom from physical presence, help in price discovery, and provide a platform for immediate sale and purchase. All these benefits are also available from an exchange providing carbon-trading facility.

Carbon registries are the records of and means to track the GHG emissions of businesses, municipalities, and other organizations working through mutual collaboration. All carbon reporting and trading programmes should have a carbon registry. Carbon registry also helps assess progress in carbon emission reduction. As it provides quantified and verified carbon offsets, it is very important for any carbon credit market. As a result, some carbon offset schemes will consist of only a registry and a portfolio of carbon sequestration projects, whereas some will add a trading platform to facilitate buying and selling of carbon offsets (Bettinger, Clutter, Siry, *et al.* 2009).

These "registries" are used for providing data for carbon-reduction initiatives to collaborators such as government and its agencies, businesses, and other interested stakeholders. Several government agencies have started registries for recording credits generated through voluntary offset projects; for example, the US Department of Energy initiated the National Voluntary Greenhouse Gases programme and CCAR, a non-profit voluntary registry for emissions reductions.

The "Centre for Resource Solutions" (Green-e Energy) operates a registry for carbon credits. Retailers such as MyClimate (a Zurich-based retailer) also operate their own registries that they sell into the voluntary market.

Stage IV: Carbon credit product consumption

Carbon credits are purchased by buyers to meet either their compliance targets or their green business objective of becoming carbon neutral. In general, carbon credits are consumed to offset one of the following four types of emissions:

- **Internal emissions**: Companies (such as HSBC), non-profit organizations, or government agencies may purchase carbon credits to counterbalance the emissions generated by their business activities. These types of emissions are also called direct emissions and account for most carbon credits purchased on the voluntary market.
- **Product life cycle emissions**: Companies may purchase carbon credits to compensate for the emissions generated when customers use their products (and thus these types of emissions are external emissions). Such carbon-neutral products carry a price premium. For example, BP offers a carbon-neutral product "ultimate grade gasoline" to its customers who are willing to pay the extra cost of purchasing carbon credits. This premium price (cost) is charged to purchase the carbon offset credits equivalent to the emissions caused by the use of gasoline by the customers. In the forestry sector, the nearest equivalent to these carbon-neutral products is "certified forest products", which is available in the market. These certified forest products are the outcome of forest management process, which are carbon intensive to some extent.
- **Event emissions**: In 2009, event organizations such as the World Forestry Congress purchased and got the sponsorship for a large number of carbon credits to make the events carbon neutral. Event sponsors pay for buying carbon credits to neutralize the carbon generated during such events.
- **Individual emissions**: Some environment-friendly individuals purchase carbon credits to make up for the carbon footprints caused by their daily activities.

GENERAL STATUS OF CARBON MARKETS

The international carbon markets transacted 8719 million tonnes (MT) of CO_2 equivalent in 2009, valued at \$144 billion. However, only a fraction of this (only 1.1% of volume and 0.3% of value) was contributed by the voluntary market, while the rest was contributed by the regulated markets (Table 8.2). Further, it is estimated that almost 10% of the total

Table 8.2 Transaction volumes and values for global carbon market (2008 and 2009)

Markets	Volume (MT CO_2 equivalent)		Value ($ million)	
	2008	*2009*	*2008*	*2009*
Voluntary OTC	57	51	420	326
CCX	69	41	307	50
Other exchanges	0.2	2	2	12
Total voluntary markets	127	94	728	387
EUETS	3,093	6,326	100,526	118,474
Primary CDM	404	211	6,511	2,678
Secondary CDM	1,072	1,055	26,277	17,543
JI	25	26	367	354
Kyoto [AAU]	23	155	276	2,003
New South Wales	31	34	183	117
RGGI	62	813	241	2,667
Alberta's SGER	3	5	34	61
Total regulated markets	4,713	8,625	134,415	143,897
Total global markets	4,840	8,719	135,143	144,284

AAU – assigned amount unit; CCX – Chicago Climate Exchange; CDM – Clean Development Mechanism; EUETS – European Union Emission Trading Scheme; JI – joint implementation; MT – million tonnes; OTC – over the counter; RGGI – regional greenhouse gas initiative; SGER – specified gas emitters regulation

Source: Adapted from Hamilton, Sjardin, Peters-Stanley, *et al.* (2010)

OTC voluntary credits was accounted by forestry projects (Hamilton, Sjardin, Peters-Stanley, *et al.* 2010).

Status of Forestry Carbon Projects

The forestry sector was the largest sector during the initial stages of OTC voluntary carbon market. However, with fast growth and diversification beginning in 2005, the share of forestry projects has gone down, but in absolute numbers it has been increasing. In 2009, the afforestation and reforestation sector comprised 10% (4.3 MT of CO_2 equivalent) of all OTC voluntary carbon transactions. The OTC voluntary carbon market facilitated 75% of all forestry carbon transactions (Hamilton, Sjardin, Peters-Stanley, *et al.* 2010) (see Figure 8.3).

In a recent survey of the afforestation and reforestation sector (Merger 2010), it has been found that of the 118 afforestation and reforestation projects for which data were collected, only 21 commenced until 2004, while another 81 were developed during 2005–10, showing a 27%

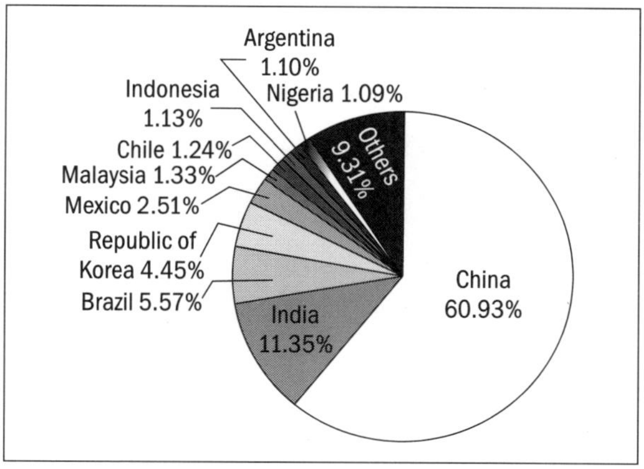

FIGURE 8.3 Expected average annual CERs from registered projects to the host countries

Note: By the CDM Executive Board as of September 2010 (total 382,430,860)

annual average increase during this period (Figure 8.4). This may be worth mentioning here that such surveys of carbon markets cannot depict the complete market picture. Thus, this survey does not include all forestry projects because of multiplicity in project registration, verification standards, and agencies involved, plus a number of carbon-trading exchanges, both in the compliance and in the voluntary markets. However, the survey captures a majority of forestry projects and thus gives a near-complete picture (see Figure 8.4).

A maximum of 30 afforestation and reforestation projects were seen in 2010 alone, but a sharp decline was witnessed in the following years

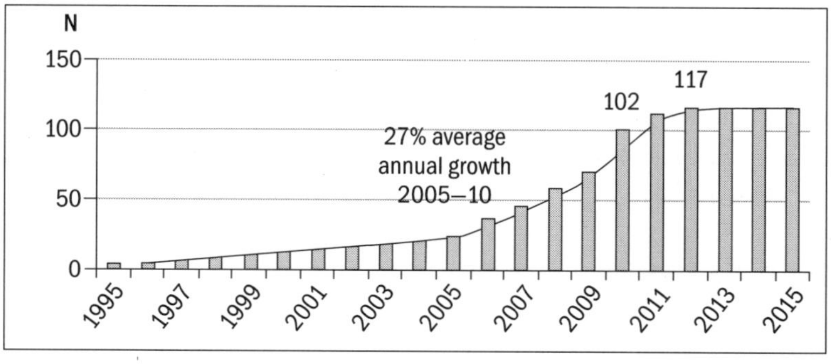

FIGURE 8.4 Commencement of cumulative afforestation and reforestation project (1995–2015)

Source: Adapted from Merger (2010)

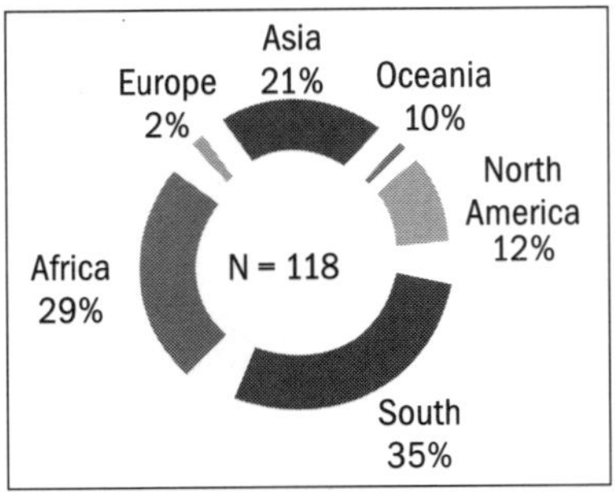

FIGURE 8.5 || Location of afforestation and reforestation projects

Source: Adapted from Merger (2010)

because of the uncertainty in the continuity of the Kyoto Protocol beyond 2012. However, with a positive signal for a legally binding emission reduction agreement emerging, the number of afforestation and reforestation projects may increase from 2012 onwards.

The aforementioned survey further reported that out of the 81 projects that came up during 2005–10, the maximum number of afforestation and reforestation projects (41) has come up or are commencing in South America, followed by 34 in Africa, 25 in Asia, and 14 in North America (see Figure 8.5).

With respect to the scale of forestry projects, the survey highlights that 14% of the forestry projects were of less than 100 ha (micro-projects), 26% between 100 ha and 1000 ha (small projects), 21% between 1000 ha and 5000 ha (mid-sized projects), 15% between 5000 ha and 15,000 ha (large projects), and the remaining 24% were of more than 15,000 ha (very large projects) (see Figure 8.6).

In terms of the total forest area by project scale, a maximum of 0.42 million ha were covered under very large projects, followed by 0.16 million ha under large projects, 59,000 ha under mid-sized projects, 15,000 ha under small projects, and only 1300 ha under micro-scale projects (Figure 8.7).

ANOTHER ESTIMATE OF FORESTRY CARBON MARKET

The total transacted forestry-related carbon projects were 67.8 MT of CO_2 equivalent up to 2008, as reported by Ecosystem Marketplace 2010 (state of the forest carbon markets 2009; taking roots and branching out). Out of this, 47 MT of CO_2 equivalent was transacted to the US

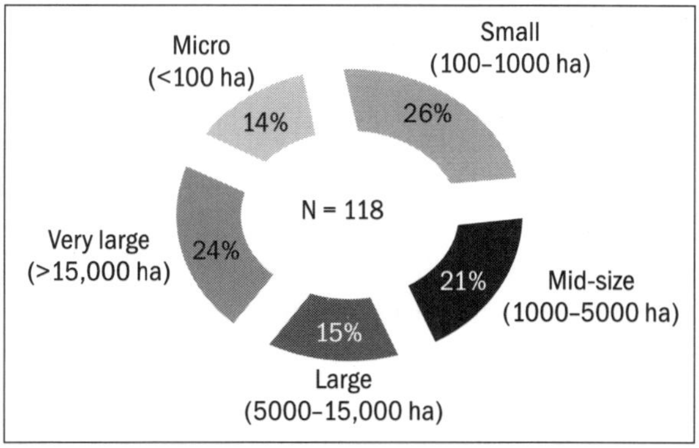

FIGURE 8.6 Scale of afforestation and reforestation projects

Source: Adapted from Merger (2010)

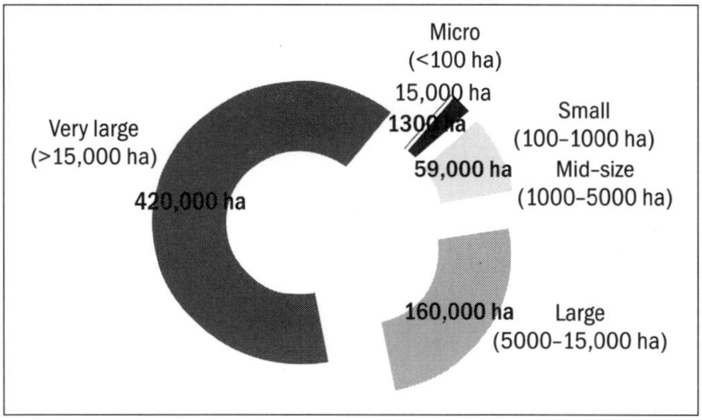

FIGURE 8.7 Total forest area by project size

Source: Adapted from Merger (2010)

electric power company AES Corp for $2 million in 1990. All other forestry carbon projects (totalling 226) up to 2008 resulted into the remaining 20.8 MT of CO_2 equivalent transactions. In 2007, the volume of forestry carbon transactions was 5.1 MT of CO_2 equivalent, which is 228% higher than that in 2006. It further increased to 5.3 MT of CO_2 equivalent in 2008. In the first two quarters of 2009, 3.7 MT of CO_2 equivalent was reported to be transacted. The global forest carbon markets were estimated at roughly $40 million in 2009.

PRICING OF CARBON CREDITS

Pricing of CERs

The price of carbon credit is determined in the carbon markets, and thus it is highly speculative. Three market systems are available for carbon product pricing: project-based market (baseline and credit), allowance market (cap and trade), and voluntary market (CIFOR 2005). The price paid by the buyers to purchase one unit of CER to offset their GHG emissions is termed the unit price of the carbon credit.

In the calculation of price settings, project proponents should include all the expenses related to marketing, project development, and financing; the services of a DOE for project validation; all the costs related to the accreditation of CERs by the CDM Executive Board; and monitoring costs (Cornland 2006). Both sellers and buyers can facilitate themselves with various options for price setting according to their exposure and speculation of the CDM market. Typically in afforestation and reforestation CDM, the risk associated with projects is much higher than that of other sector's CDM projects due to the temporal nature of carbon sequestration, which requires adequate conservation measures. A project proponent can sell CERs at any stage of the project development according to their financial needs, even on the basis of "Project Idea Note" or "project proposal".

The CERs issued after validation and registration can be directly traded in the market on the current market price, which does not involve complex procedure; however, selling of CERs before registration involves various risks and vulnerabilities such as non-approval and natural disasters. In this scenario, the selling price of CERs or payment schedule can be fixed in the following ways (Neeff and Henders 2007):

- A fixed price for each unit transaction of CERs traded between the parties.
- A percentage of the market price of a EU emission allowance (EUA) (as a reference) for each delivered CER before an agreed date.
- An indexed price with a floor and a ceiling.

This price setting with various arrangements involves a risk for both the seller and the buyer. With the option of fixed price in the agreement regardless of the market development, the price of CERs remains the same and subsequently the project proponents are devoid of advantageous market situations. On the other hand, such an arrangement carries the advantage that certain future cash-flow projections can be made (Neeff and Henders 2007).

FOREST CARBON PRICE AND ITS IMPLICATIONS FOR FOREST MANAGEMENT

At present, the economic value of above- or below-ground forest carbon in carbon markets is relatively low, and as a result, the current impact of carbon sequestration on forest management planning is relatively low. A reason for this is that quantifying and verifying carbon offsets is a lengthy and expensive process, and low carbon values do not justify very active participation in the carbon offset trade. Further, at this point in time, all carbon-trading programmes in the USA are voluntary; thus, landowners are encouraged, not forced, to participate. Should there eventually arise legally binding regulations that limit emissions, carbon offset values may increase substantially, thereby justifying more attention from forest managers and landowners (Bettinger, Clutter, Siry, *et al.* 2009).

Pricing of VERs

The prices of carbon retail offsets vary widely. Carbon-neutral websites sell offsets ranging from $5 to $25 per tonne, averaging about $10 per tonne in their offerings. Price variety and competition might appear to be good for both consumers and market expansion; however, with respect to environmental integrity, low offset prices can be a double-edged sword.

The prevailing price of carbon in the US voluntary markets varies from less than $2 under RGGI to $4–7 in California; however, according to the Global Change Associates, the right price band should be between $30 and $50 per tonne to attract investment into this nascent carbon industry.

The price of vintage 2009 Carbon Action Reserve (CAR) VERs was $2.68 in the New York Mercantile Exchange trade on Friday, 17 December 2010, after the permission given by California's ARB. A shortage of eligible CAR credits would help in further price hike in 2011 and trigger new project activity. In California, the first forward trade of carbon emission allowances has been reported to take place at $11.50 per tonne.

Pricing of EUAs

The price movement for EUA between 2007 and 2010 (August) was in the range of €31 (August 2008, just before the financial crisis in the US economy) to €8 (at the end of 2008); it remained more or less stable at about €13–15 per EUA during 2009 till date. The price of EUAs in the EU carbon market remained close to €14 on the European Climate Exchange in December 2010.

PROMOTIONS FOR MARKETING OF CARBON CREDITS

Carbon markets are influenced by the perceived risk associated with project and are inversely proportional to the price of CERs offered by the seller. "The higher the seller's perception of investment risk, the more urgent it is to reduce that risk by obtaining the commitment of a buyer to the purchase and a price at least, and potentially to other terms such as the sharing of project development costs and the provision of support in the form of access to expertise" (Cornland 2006). Thus, there remain high demands for a trustworthy project and can deliver the CERs, which were proposed both on time and with adequate quantity. These types of projects are less risky and known as bankable projects.

Thus, a bankable CDM project with a strong proponent (seller) will run the project smoothly and deliver the full anticipated quantity of CERs on schedule. Such projects require certainty, devoid of any fuss and protest from project stakeholders. Also a bankable project should meet the following criteria (Cornland 2006):

- It should not run into financial difficulty.
- It should get through the regulatory process without much difficulty.
- It should not fail or stumble technically.
- It should not lead to contractual default on the part of the seller.

Therefore, a good promotional strategy for carbon credits should highlight the information that makes a project bankable and trustworthy and should be able to couple it with the preferences of buyers. When a CDM project comes into the market, buyers initially scrutinize the project to acquire information about its eligibility, to gauge the accuracy of the proposed carbon sequestration by the project, and to know about its profitability. They normally look for technical aspects of the project such as sustainability, additionality, and economic viability, as well as the proficiency of project proponents in carrying out the project. Buyers from the government or public sector are more concerned about issues of sustainable development or inclusion of environmental benefits from the project. Buyers particularly interested in afforestation and reforestation CDM projects may enquire about the implication of biotic pressure in and around the project boundary.

The producers of forestry carbon projects in India may normally take a longer period of agreement (a minimum of five years duration) because of the non-permanence nature of forestry projects. The agreement is done only when the buyer is satisfied with all the adequacies of the forestry projects. Hence, the seller should form a strategy highlighting all the eligible features of the concerned project.

Buyers' Preference of Carbon Credits from Forestry Projects

Buyers (mostly governments and banks) prefer projects that generate both environmental and socio-economic benefits simultaneously. The involvement of local communities and other stakeholders in the project can deliver credible benefits to the rural poor in the developing countries, leading to more integrated approach to sustainable development. These types of projects are more appealing because they furnish dual goals of environment restoration, such as biodiversity conservation, protection of soils, and combating land degradation and desertification. Certification of the project design according to the CDM quality standards is one way of demonstrating sustainability co-benefits and may thus ultimately contribute to market success. One such standard is the Climate, Community, and Biodiversity Standards, which evaluate land-based carbon mitigation projects in the early stages of development on the following four parameters (Neeff and Henders 2007):

- Is the project addressing climate change?
- Is it supporting local communities?
- Is it resulting into conservation of biodiversity?
- What is innovativeness and vulnerability to market risk?

Buyers of Carbon Credits/Products

The potential buyers of CERs can broadly be classified into three categories: government bodies, associates of government, and private organizations.

Government Bodies and their Associations

Europe

The representatives of several European governments regularly declare about considering the purchase of carbon credits from the forest sector. Spain and Ireland have recently invested into the second tranche of the BioCarbon Fund (BioCF) (World Bank 2007). Spanish sources are continuously funding for backstopping forestry CDM as an engine for rural development. Also, new carbon funds are being designed to purchase carbon credits from forestry CDM.[3] Forestry CDM market transactions will likely expand, as the project pipelines of forestry CDM projects are strengthening and the governments are stepping up their purchase strategies (Neeff, Lisa, Imme, *et al.* 2007). The decision of the sixteenth meeting of the Conference of the Parties (CoP) at Cancun for continuation of the LULUCF work programme, as part of the Kyoto Protocol phase II, provides enough evidence that the carbon market for forestry projects would continue to expand and further evolve.

Canada

Although Canada is obliged to reduce its emissions 6% below the 1990 level under the Kyoto Protocol, it emits 27% more than the target at present. At this rate, emissions could increase up to 50% above the country's target by 2012. Canada has put itself into a difficult position by neglecting effective policies for reducing emissions. In February 2007, the Canadian Ministry of Environment declared that Canada would not stick to its Kyoto target; rather, it would aim at reducing its annual GHG emissions by approximately 20% (that is, 150 MT of CO_2 equivalent) below its current level by 2020 (instead of reducing 300 MT of CO_2 equivalent by 2012, as required to achieve the Kyoto target), according to a new climate plan.

Japan

Under the Kyoto Protocol, Japan is obliged to reduce emissions by 6% below the 1990 levels in 2012; on the contrary, its emissions have increased by 8.1% above the 1990 levels. The Japanese government relies on carbon credits for meeting the Kyoto goals. Hence, this increased emission leads to a great potential demand for Kyoto-compliant carbon

[3] Details available at <www.proyectoforma.com>

credits. Thus, the budget of carbon credit has almost been tripled in 2007 from 6.4 million credits bought in 2006. Presently, the Japan government's carbon credit purchase programme, managed by the New Energy and Industrial Technology Development Organization, excludes the forest sector (Neeff, Lisa, Imme, *et al.* 2007).

World Bank

The World Bank remains a major buyer of CERs generated by the afforestation and reforestation CDM through its mechanism of carbon funds. The multilateral funds for carbon projects are overwhelmingly dominated by the World Bank, which facilitates eight carbon funds: prototype carbon fund, Italian Carbon Fund, the Netherlands Clean Development Facility, Community Development Carbon Fund, BioCF, the Netherlands European Carbon Facility, Spanish Carbon Fund, and Danish Carbon Fund.

Canada, Finland, Italy, Netherlands, Norway, Sweden, and other European countries are some of the most active contributors to these carbons funds. Money in these funds is being channelized from industrialized countries and companies to host countries and communities, to foster carbon projects and produce carbon credits. Fund is also used for pre-financing for project identification and preparation activities such as capacity building, outreach, and research, leading to the creation of supportive project approval systems in host countries. Therefore, these funds are considered an alternative for both pre-project financing and selling of carbon credits (Pearson, Sarah, and Sandra 2006).

Private buyers/brokers

Scattered organizations, brokers, and firms that seek carbon-neutral status or want to cut down emissions due to legal regulations imposed can be categorized into this section. Private buyers/brokers want to use them to offset their own GHG emissions. The broker organizations are interested mainly because they want to accumulate the CERs for future prospects, speculating the supply crunch in the future.

FUTURE OF GLOBAL CARBON MARKET

The earnings from CERs/VERs provide an extra income for the forestry project proponents; marketing aspects are the ways to ensure or secure that income. These provide a leverage to promote the beneficial aspects of the project and thus generate a demand for the carbon product from the project, consequently helping in fetching the revenues from carbon credits.

Despite being small and fragmented, the voluntary market is growing rapidly. Thus, it provides a respite for the forestry carbon projects other than the compliance markets through afforestation and reforestation. Policy developments in the regulatory sector, such as the ratification of the Kyoto Protocol and the EUETS, appear to have boosted carbon markets along with the motivation of voluntary markets. Policy developments have raised the profile of climate change in the media, thereby fuelling the sentiment that individuals and organizations should take more responsibility for their impact on the climate. The current regulatory regimes in Europe and Australia and the ones planned elsewhere cover only large emitters; hence, companies, organizations, and individuals have plenty of scope to be active in the carbon market. This will expand the carbon markets in general and forestry projects in particular.

Although at the Cancun CoP-16 of 2010, an agreement on the Kyoto Protocol phase II could not be reached, the declaration included the intention to conclude the deal possibly at Durban. The survey of key stakeholders in the carbon market predicts that there will be a huge growth in carbon market, benefiting businesses in both the developed and the developing countries (Box 8.5).

Despite the fact that the global carbon market is expanding, even less than 0.1% of carbon capture is based on removing the existing carbon dioxide out of the atmosphere. The global forest carbon markets were roughly $40 million in 2009. It is very important to engage in developing a low-carbon economy; at the same time, it is equally important to remove the existing carbon dioxide out of the atmosphere, as this is a key to mitigating climate change. Thus, forestry needs to play a major role in global GHG mitigation efforts to achieve emission reductions; the forestry sector increasingly gains public attention and market acceptance within this arena (EcoSecurities 2010).

At the global level, the Cancun conference of 2011 called for "deep cuts" in emissions, establishing a Green Climate Fund to help poor countries tackle global warming (reaching up to $100 billion by 2020), a scheme for compensating countries that desist from felling their forests (*Economic Times,* 11 December 2010).

At the regional level, California's cap-and-trade scheme, cleared in December 2010, is emerging as a critical market for forest-based

BOX 8.5 The survey of carbon market players about its future scope

According to the organization Point Carbon (2 September 2010), an Internet survey on 70 respondents—constituting analysts, traders, brokers, academicians, and lawyers—highlighted that 50% expect the value of international carbon trade to reach between $500 billion and $1 trillion by 2020. However, the risk in carbon markets looms large over the post-Kyoto Protocol period. The developed and developing countries are still at odds on issues of funding mechanism, mitigation and adaptation strategies in poor countries, and monitoring greenhouse gas emission cuts in developed and major emerging economies. Forty per cent of respondents in the survey expect that a UN climate deal would be reached by 2012, and 53% expect that a post-2012 agreement would continue to be legally binding. About 30% of the respondents believe that Clean Development Mechanism (CDM), joint implementation, and trade in assigned amount units would continue beyond 2012, and 51% believe that Asia and about 40% believe that the USA would be the most attractive region for investments to generate carbon credits in the developing and developed regions, respectively. Thirty six per cent believe that the European Union would go for a 20%–30% emission cut for 2020 whether or not the developed countries make similar commitments post-2012.

Source: Point Carbon (2010)

offset credits. The rules of the forestry offset protocol would make carbon credits from afforestation and reforestation, improved forest management, and avoided conversion activity eligible for compliance in the California trading scheme. Offset projects from Canada and Mexico and project-based offsetting, including avoided deforestation, or Reducing Emissions from Deforestation and forest Degradation (REDD), in tropical developing countries such as Brazil and Indonesia may also be eligible in the future.

Certification of carbon credits, which is also a new trend, is emerging; this would allow market to differentiate between products and to introduce price premiums for certain types of projects. This certification will be able to increase both the supply and the demand of carbon projects as confidence in the quality and quantity increases (Lichtenfeld 2007).

FORESTRY CARBON PROJECTS—THEIR STATUS AND FUTURE

To date, the carbon market could be able to generate marginal finance for REDD+ activities only. The World Bank Report on Mapping Carbon Pricing Initiatives from May 2013 speaks of an unchanged result of 7 $MtCO_2e$ in 2011 and 2012. The Bloomberg Report "Maneuvering the Mosaic–State of the Voluntary Carbon Markets 2013" assesses a similar situation. The voluntary market, in which forest projects play virtually a significant role among all possible projects, naturally brings only modest results in terms of quantities. The prices paid for certificates in 2013 (US$7 or 8—and recently US$5 in Norway) illustrate a low pricing level that stems from small quantities and lack of demand. However for the future, the Californian Cap and Trade Programme promises take up capacity of up to 80 $MtCO_2e$ in the period to 2020 (CMR 2013).

The compliance carbon market for forestry projects, that is, the CDM market, offers a comparatively sober picture. Very few CDM forestry projects have been registered so far. This was and still is not just due to the lack of demand for certificates, but also due to the barriers involved in the process. Thus, the contributions already made to climate change mitigation and those expected to be made, through forestry are minimal so far. However, it is a fact that global annual emission of around 3000 $MtCO_2e$ (Harris, Brown, Hagen, *et al.* 2013) caused by deforestation will result in complete deforestation within a generation in some countries and regions (CMR 2013).

In theory, the carbon market could be an important source of funding. Till date there have been marginal finance amounts. This alone is reason enough for further efforts to bring together market finance and carbon forestry (CMR 2013).

However, this option has attracted much criticism right from the beginning. Apart from objections raised by market opponents, whose position has somewhat weakened since the first CDM debate about 15 years ago, there are still a range of technical problems, which by their very nature should be treated as solvable problems. These involve the issues of ensuring permanence in carbon sequestration, preventing leakage, and establishing good governance at both national and multinational levels (CMR 2013) and deserve to be given serious consideration.

9

LIVELIHOOD ISSUES IN FORESTRY CARBON PROJECTS

A David and H S Gupta

CLIMATE CHANGE AND LIVELIHOOD

In the past, co-benefits of forestry carbon projects have received less attention as compared to their emission reduction and carbon sequestration aspects. However, the spotlight on livelihood and sustainability has rightly been turned on, with such co-benefits seen as being real, "additional", and, possibly, measurable. These concerns arose out of the need to not just eliminate the possibilities of projects being detrimental to the local communities' well-being but, moving beyond, to provide positive inputs. Offset buyers would then base their decisions not only on the quantity of carbon offsets but also on the "quality", as reflected in the livelihood and sustainability aspects (EcoSecurities 2010).

FOREST CARBON PROJECTS AND THEIR BENEFITS FOR LOCAL PEOPLE

People living in villages within forest areas or on the forest fringes traditionally depend on forest-based resources for sustenance and livelihoods. The villagers fulfil their demand for firewood, fodder, and thatch from the adjacent forest. The most economically deprived populations, however, are almost entirely dependent on forests for their livelihoods, which range from cattle grazing in forest areas, gathering fruits and other non-timber forest products (NTFPs), and deriving a larger part of their income from wage labour. In view of the close linkages people traditionally have had with forests, extensive reforestation, preventing deforestation, and other land-based emission reduction activities render local people highly sensitive to changes in land-use options.

On the one hand, many rural communities are eager to work in forestry carbon projects for extra income, job, and other social benefits, along with offset buyers interested in emission reduction and, at the same time, helping the local people. On the other hand, sceptics are wary that these projects may do more harm than good and can result in the loss of (uncompensated) traditional livelihoods or resource-access rights of the poorer community members. Accurately projecting and then measuring both the positive and negative impacts of a project not only is a moral imperative but also has important commercial implications (Richards and Panfil 2010). The co-benefits arising out of afforestation projects in general and forest carbon projects in particular are summarized as follows:

A good forestry carbon project may provide increased access to forest resource base and forestry products such as NTFPs, medicinal plants, gums and resins, fuelwood, and timber. In many large forested Indian states, which also support large populations, collection of *tendu* (*Diospyros melanoxylon*) leaves, along with honey and mahua, provides a definite supplemental income to the forest-based communities. Many communities rely on these forest products for meeting their subsistence needs and generating extra income by selling them.

Although forests often provide needed goods, services, and supplemental income, sound forest management may not be profitable in all situations. In such scenarios, payment from carbon credits increases the value of forests, relative to other conventional land uses. Thus, these additional carbon payments can be used as a tool to finance activities required to achieve sustainable management regimes. The additional revenue thus generated by carbon benefits can be well utilized to avail the services of good nursery and improved seeds (CIFOR 2001).

Forestry carbon projects can amplify the efficacy of the nearby watershed area, thereby increasing both surface and ground water levels. This enhanced level of water can be used for both drinking and irrigation purposes, thus providing an additional water supply for the generation of crops and yield, which are critical for livelihood. A recent example is the Mid-Himalayan Watershed project in Himachal Pradesh, which is one of the largest Clean Development Mechanism (CDM) projects of its category.

Plantation can prevent soil erosion and landslides through sedimentation of water. Forestry projects can restore millions of hectares of degraded lands in the developing countries. This will further control soil erosion and nutrients depletion and will restore the fertility and productivity of the agricultural fields, subsequently help maintain livelihoods if not enhancing them.

Improvement in the social network and capital of the community through joint forest management (JFM) activities is another area where forestry contributes to livelihoods and well-being. Active participation in forestry projects provides a means for local communities to exercise their indigenous knowledge and to further strengthen their expertise in effective management strategies. Forestry development projects have often strengthened local institutions by involving communities in decision-making.

Forestry carbon projects provide greater scope for the development of forest-based small business enterprises or production units. The by-products of these forestry projects (such as silk or gum) can be used as input materials for other industrial purposes. Leaves from mulberry plantation can be used in silk production units, and honey can be directly sold to food-processing industries. *Tendu* leaves can be used for making *beedi* (Indian cigarettes).

Forestry carbon projects help in sustainable management of forest resources. In addition, these projects can also help establish more sustainable patterns of energy consumption by enhancing the energy efficiency of forest product use through processing plants. For instance, forestry and renewable energy projects have significantly enhanced biomass fuel resources and improved energy efficiency by introducing advance and energy-efficient cooking stoves and charcoal kilns (CIFOR 2001).

Additionally, opportunities for livelihood from forestry activity can be increased through recreational activities such as ecotourism or through conservation of wildlife. In the ensuing paragraphs, specifically consider the livelihood dimensions of some forest-based carbon projects.

Large-scale Industrial Pulp or Timber Plantations

Industrial plantations require skilled and unskilled labour for project development and for activities related to plantation and its maintenance.

Such plantations require more labour as compared to agriculture. These plantation activities can help local inhabitants through employment and income generation. The requirement for labour decreases after the initial plantation phase; however, if the harvesting pattern of trees is cyclic as planned, then the labour requirement will remain constant (Smith and Scherr 2002). Sometimes communities are not allowed to access large-scale forestry projects or protected forests; this poses a threat to their livelihood and sustainable development opportunities because their lifestyle is dependent on the degraded land that has been used to develop the forestry project.

If any biodiversity conservation projects prevent harvesting of forest products, the access to locally important sources of forest products or income is restricted. Previously, the project area was maintained by the villagers and now they lost their jobs because of the advent of project (Smith and Scherr 2002). Consequently, project proponents may strive to develop a participatory model in which both communities and forests can prosper, along with ensuring that local inhabitants should not be the sufferers.

Community Forestry Plantations

In community forestry, forests are maintained on community lands to provide products such as gum, resins, fruits, and latex. These products supplement the income of all stakeholders, including nearby communities. Extraction of these forest-based products results into minimal impact on the existing carbon stocks. Community forestry, in particular, can be termed a successful programme primarily because of social inclusion into the projects and secondly because of its typical benefit-sharing mechanism that allows its stakeholders to access the products and services of the projects in a sustainable manner. Community forestry promotes social inclusion, poverty reduction, and equity within the communities by enhancing the livelihood opportunities. After value addition, when forest produce and services are diverted to the market, they subsequently generate surplus revenue for the local inhabitants (Chapagain and Banjade 2009).

The tamarind project in southern India with Plan Vivo support is an example of agroforestry and bioenergy project. Under this innovative project, small farmers with holdings of about 2 ha have organized and entered into a carbon sale agreement with "Future Forests", an international company based in the UK. The mango and tamarind plantations have been increased to more than 2 ha of land belonging to a small farmer and 18 tonnes of carbon is expected to be fixed over six years. The fixed carbon has been agreed to be sold at the rate of $10 per tonne of carbon, equivalent to ₹8640. The company should pay this amount to the farmer in five instalments, starting from the second year: 50% (₹4320) as the first instalment in the second year, 20% as the second instalment in the third year, and the remaining 30% in the remaining three instalments in the fourth, fifth, and sixth years. Thus, in addition to the income generated by the sale of fruits and other products, farmers will get an extra income from the sale of carbon (adapted from Satyanarayana 2004).

Agroforestry Plantations

Agroforests can be grown on fallow or crop land and mainly include valuable species such as cinnamon and coffee; thus, these are very useful in generating revenues for farmers (Smith and Scherr 2002). Agroforestry can be incorporated in forestry projects where the benefit to local communities is preferred to the carbon credits because they may not generate adequate carbon credits required to make the project profitable. Moreover, such projects improve biodiversity and add value to the ecosystem services that help the communities residing near the forests.

Forestry Regeneration and Rehabilitation Project

Forests that suffer damage due to commercial exploitation and overuse may be allowed to be restored by the communities to regenerate the linkages of services and benefits from forests. These restoration processes can directly benefit the local communities by providing them the lost sources of natural resources and NTFPs. Forest regeneration activities can also provide direct employment opportunities along with the restoration of broken link between communities and forests. Thus, the stake of community, if created, may help in sustained conservation of projects.

FOREST CARBON PROJECTS AND SUSTAINABLE LIVELIHOODS: PROVISIONS AND THE ROLE OF STANDARDS

The sustainable livelihood model (as described by the Department for International Development) aims at developing or enhancing five types of capitals: natural capital (water, forest, and land), physical capital (farm tools, roads, and other infrastructure), financial capital (savings, incomes, and credit availability), human capital (health and education), and social capital (group formation, social cohesion, and kinship) (Smith and Scherr 2002). The inclusion of "sustainable development" (Article 12.2 of the Kyoto Protocol) into the definition of the afforestation and reforestation CDM projects recognizes poverty reduction as a legitimate component in the role of such projects not only in carbon sequestration but also in livelihood support. Even though sustainable development as a concept has not been elaborated in the Kyoto Protocol, a general understanding could be that it aims at enhancing the local resources base without compromising on the future availability. According to this principle, any project should incorporate participatory approach, flexibility, and adaptive responsive mechanisms to be sustainable. Forestry carbon projects with sustainable development approach, therefore, not only aim at utilizing natural capital such as forests but also enhance the other capitals as well. For instance, forest products and medicinal plants contribute to human capital, sales of timber and carbon credits provide financial capital, and approaches such as JFM provide a basis for social capital generation. Thus, CDM forestry carbon projects can deliver the much needed co-benefits essential for sustainable development, by encompassing the needs of local inhabitants.

A later development in the areas of climate change regulations is that of recognition of Reducing Emissions from Deforestation and forest Degradation (REDD) and REDD+ in creating incentives for conserving and managing forests in the developing countries. REDD+ envisages incentives and monetary benefits to the local communities on the basis of the value of carbon stored and enhanced in soil and in the standing biomass. It is estimated that a REDD+ programme for India could capture more than 1 billion tonnes of additional CO_2 over the next three decades and provide more than \$3 billion as carbon service incentives under REDD+. India is committed towards transferring monetary benefits from REDD+ to local, forest-dependent, forest-dwelling, and tribal communities (MoEF 2010).

A series of principles evolving from the consultations of several stakeholders are included under the Social and Environmental Standards (SES), which are as follows:

- Rights to lands, territories, and resources are recognized by the REDD+ programme.

- The benefits of REDD+ programme are shared equitably among all rights holders and stakeholders.
- The REDD+ programme improves long-term livelihood security and well-being of indigenous peoples and local communities, with special attention to the most vulnerable people.
- The REDD+ programme contributes to broader sustainable development, respect and protection of human rights, and good governance objectives.
- The REDD+ programme maintains and enhances biodiversity and ecosystem services.
- All relevant rights holders and stakeholders participate fully and effectively in the REDD+ programme.
- All rights holders and stakeholders have timely access to appropriate and accurate information to enable informed decision-making and good governance of the REDD+ programme.
- The REDD+ programme complies with applicable local and national laws and international treaties, conventions, and other instruments.

Apart from the afforestation and reforestation projects under the CDM, a majority of forestry carbon projects are receiving attention from the voluntary carbon offset market. The arrangements under this voluntary market include a wide range of exchanges, brokers, and buyers who make direct purchases, and are the only sales outlet for carbon credits generated out of REDD+ (Jagger, Sills, Lawlor, *et al.* 2010). The Verified Carbon Standard (VCS) and the Climate, Community, and Biodiversity Standards (CCB Standards) are the two most popular standards concerned with projects demonstrating co-benefits for biodiversity and local communities. The popularity and market premium for CCB certification (EcoSecurities 2010) confirm the importance of understanding welfare outcomes of REDD+ projects.

Buyers are interested in guarantees to the effect that credits being purchased by them not only reduce the net carbon emissions but also do not negatively impact biodiversity and local livelihoods. Indeed, what will attract buyers are the proposition and a firm assurance from the prospective sellers or projects that local people will positively benefit from the project activities and outcomes. About a dozen standards and certification systems (Appendix 5) have been developed to provide such assurances, but only some of these certify REDD+ projects and address questions about additionality, leakage, and permanence of REDD+, as well as looking at impacts on local people and the environment. Some of the significant standards pertaining to the livelihood dimensions are discussed below.

CCB Standards

The CCB Standards differentiate quality projects that incorporate concerns with regard to the rights of local people, generate benefits, as well as conserve biodiversity. The Climate, Community, and Biodiversity Alliance (CCBA) maintains a registry of projects that have been certified to its standards. However, they do not issue emission credits or approve specific methodologies (Richards and Panfil 2010). CCBA certification is highly desirable for prospective users regardless of the scope for sales of credits. In a recent market survey, many buyers were found to be willing to pay a price premium for projects with both VCS and CCB certification (EcoSecurities 2010). The CCBA is also working to pilot the REDD+ SES for governmental REDD+ programmes to demonstrate their social and environmental "co-benefits". The SES is operationalized from the concept and indicators perspective, and the three concepts pertaining to the community aspect are given in Box 9.1.

Plan Vivo Standards

A range of land-use projects such as afforestation and reforestation, agroforestry, restoration, improved forest management, and REDD are accepted by Plan Vivo standards. Such projects are developed with small-scale farmers, based on a broader "Plan Vivo" livelihood strategy. Generally, no definite methodologies are provided by Plan Vivo, which is different from other standards. Instead, in this case, each project must devise its own technical specification adapted to the specific realities of the project and reviewed by external experts. Projects are issued Plan Vivo Certificates on an ex ante basis to ensure sufficient start-up funds for farmers (although payments are disbursed gradually).[1]

SOCIALCARBON Standard

SOCIALCARBON, a standard comparable to CCB, is designed to demonstrate social and environmental co-benefits of carbon offset projects and also to help increase active participation of local stakeholders. Generally, this standard is used in conjunction with another approved carbon-accounting standard (such as VCS or CDM). Hence,

[1] Details available at <www.planvivo.org>

BOX 9.1 ***Community concepts and indicators***

1. **Net positive community impacts**: The project must generate net positive impacts on the social and economic well-being of communities and ensure that costs and benefits are equitably shared among community members and constituent groups during the project lifetime.
2. **Off-site stakeholder impacts**: The project proponents must evaluate and mitigate any possible social and economic impacts that could result in the decreased social and economic well-being of the main stakeholders living outside the project zone resulting from project activities. Project activities should at least "do no harm" to the well-being of off-site stakeholders.
3. **Community impact monitoring**: The project proponents must have an initial monitoring plan to quantify and document changes in social and economic well-being resulting from the project activities (for communities and other stakeholders). The monitoring plan must indicate which communities and other stakeholders will be monitored, and must identify the types of measurements, the sampling method, and the frequency of measurement (CCBA 2008).

Community Gold Standard

This Gold Level Exceptional Community Benefits criterion recognizes project approaches that are explicitly pro-poor in terms of targeting benefits to globally poorer communities and the poorer, more vulnerable households and individuals within them (CCBA 2008). Projects desirous of obtaining the Gold Standard would then have demonstrated contribution to reducing poverty and enhancing the sustainable livelihoods of the poorest sections of the community. There are several other standards designed to provide additional coverage over and above carbon-accounting standards, such as Social Carbon, although these have much more limited coverage. Finally, there are also standards that seek to cover both carbon accounting and social benefits, such as Plan Vivo.

SOCIALCARBON does not establish any project type or methodology of its own. As in the case of CCB Standards, it also does not issue carbon credits.[2]

[2] Details available at <www.socialcarbon.org>

Voluntary Carbon Standard

Another leading standard is the Verified Carbon Standard, formerly known as the Voluntary Carbon Standard. In 2009, it was used by more than a third of all credits traded in the voluntary market (Hamilton, Sjardin, Peters-Stanley, *et al.* 2010). REDD+ projects intending to certify their credits to the VCS are first required to develop and obtain "double approval" of methodologies for establishing baselines, adjusting for leakages, and monitoring, reporting, and verifying land activities and emissions. This stringent process has been a significant hurdle to overcome and, as of August 2010, only one methodology has been approved for avoided planned deforestation. However, several others are currently under review.[3]

GUIDELINES FOR FORESTRY CARBON PROJECTS TO ADDRESS LIVELIHOOD ISSUES

An ideally designed forestry carbon project will provide an integrated approach that will include community participation, ensure no negative or minimal impacts on the existing resources of the community, and at the same time increase the overall livelihood opportunities along with the targeted carbon offsets (Nigel, Maria, Vargas, *et al.* 2002). Some of the steps to be followed to make communities able to provide better livelihood prospects are as follows:

- Explicitly include a broad range of forest management and agroforestry activities in the programmes on climate change.
- Provide a broad scope for strong local stakeholder and community participation.
- Implement social impact assessments for all forestry carbon projects.
- Strengthen capacity at the local, national, and international levels.
- Provide incentives for projects with multiple benefits.
- Reduce transaction costs of community-based projects.

Hence, it is evident that forestry carbon projects can provide additional benefits to the communities associated with the projects. These projects can enhance both income and livelihood opportunities. Thus, land-use change and forestry carbon projects with significant livelihood benefits

[3] Details available at <www.v-c-s.org>

are both possible and desirable to achieve the dual goals of CDM, climate change mitigation, and sustainable development. Therefore, forestry carbon projects should be designed and managed with an aim to increase the well-being of the poor. Incorporating provisions such as those proposed here would reduce the risks to local people, increase the appeal to private investors, and increase the chances of success for forestry carbon projects.

10

POLICY AND LEGAL ASPECTS OF FORESTRY CARBON PROJECTS

H S Gupta

BACKGROUND

Forestry carbon projects, including afforestation and reforestation projects under the Clean Development Mechanism (CDM), are supposed to be far more beneficial to the society and local inhabitants, because they develop new long-duration carbon sink, as well as add value to the barren/non-forested/degraded land, create new forests, and thus pave the way for income-generation activities for the forest-based communities. Box 10.1 explains the legal basis of CDM and joint implementation. Thus, communities residing near such projects can expect to get benefitted by revenue generation through both carbon credits and the ecological service they draw from the forest. But if basic policies and legal aspects of any project formulation and implementation are not given due consideration, it may result into a very complex situation. Hence, this chapter attempts to provide an understanding of policy and laws related to different aspects of a forestry carbon project, at different phases.

THE BROAD PICTURE THROUGH POLICY AND LEGAL FRAMEWORK AND FORESTRY CARBON PROJECTS

It is important to assess the broader political and policy/regulatory environment in which the forestry sector operates, particularly with regard to the implementation of the Kyoto Protocol or the voluntary markets. The forestry sector has little control over policies/regulations on which the opportunities and constraints of the forestry sector are

BOX 10.1 ***Legal basis of Clean Development Mechanism and Joint Implementation***

The Clean Development Mechanism (CDM) is based on Article 12 of the Kyoto Protocol. Article 12.2 sets out two equally weighted objectives: (1) to assist investor countries in achieving compliance with their Kyoto Protocol commitments and (2) to assist host countries in their efforts to achieve sustainable development.

The modalities and implementation procedure for CDM projects are set out in the Marrakech Accords. These are designed to ensure that a real contribution is made towards achieving sustainable development in the host countries and combating climate change. In contrast to the rules of joint implementation, those for the CDM provisionally entered into force in 2001 by way of a Conference of the Parties (CoP) decision made in Marrakesh (known as the "prompt start"). In the following years, the CoP has adopted additional guidelines for CDM implementation along with modalities and procedures specifically applicable to afforestation and reforestation projects. Following the Kyoto Protocol's entry into force, the CoP/MOP adopted all the CoP's provisional decisions regarding CDM and issued further guidelines (decisions 3–8/CMP.1). The CDM Executive Board is responsible for drawing up detailed rules and guidelines and for monitoring CDM projects.

Hence, as part of their response to climate change and despite their political stance towards the Kyoto Protocol, governments need to make complex policy decisions keeping in mind their varied emphasis on maintaining existing carbon reservoirs, enhancing carbon, and enlarging carbon stocks. In many respects, climate change can be alleviated by the conservation of remaining forest stocks, which is a very important as well as the most cost-effective activity. Except Russia, the countries that have not signed the Kyoto Protocol and have less incentive to maintain forest stocks are most likely to be affected by an increase in deforestation. Therefore, notwithstanding the policies and instruments developed under the protocol, national, bilateral, and multilateral forest policies may reflect the objectives of the protocol and will be adjusted accordingly.

Source: Walker, Pearson, Munishir, *et al.* (2008)

highly dependent. This sector is vulnerable to the influences of other sectors' policies related to agriculture, rural poverty, and energy. Box 10.2 explains how government can play a practical role in this regard. Despite its cyclical nature and substantial opportunity costs in comparison to other important sectors such as agriculture, investment into forestry sector has always been low. Consequently, in the so-called developed countries, economically viable forestry has to depend on supportive policies always; as a consequence, the sector has become particularly vulnerable to policy distortions or adjustments. The forestry sector is dependent on long-term policy instead of short-term market signals; this makes it highly risky to respond positively to incentives. Consequently, the sector is likely to be marginalized in the debate over tangible emission reductions and the flows of technology and finance that the non-Annex 1 countries wish to enhance. In case the forestry sector plays a substantial role alongside the energy sector in mitigating climate change, equivalence in sector policy would likely to favour investment from companies and investors involved in each sector.[1]

The urgent need to control climate change can be supported by maintaining and enlarging the global forest resource. However, the burden of increased management may result in the sector not participating fully in the efforts to mitigate and alleviate climate change impacts. Furthermore, considering the fact that the incentive framework established under the Kyoto Protocol was to compensate for a market failure, it is obvious that the success of the protocol's implementation will greatly depend on innovative policy interventions (Walker, Pearson, Munishir, *et al.* 2008).

Yet, activities such as enrichment planting, establishment of additional protected forest areas, and development of non-timber forest products, which support carbon sink, storage, and sequestration activities, are not preferred by the private sector due to predicted low profitability. Hence, the public sector intends to develop a range of policies, including subsidies, to add to the possible value of carbon credits; however, only in rare instances will this be a sufficient incentive to mobilize additional investment. Development of comprehensive land-use plans and forestry sector strategies to make adequate provisions for the private sector and increase social stakeholder participation is required (Walker, Pearson, Munishir, *et al.* 2008). The difficulties encountered during the

[1] Details available at <www.forest-trends.org>

BOX 10.2 ***How can governments assist with proactive policies and regulations?***

Despite great opportunities for project development, it has been hindered by a number of regulatory barriers in the forestry sector. Governments can take the following potential actions to reduce the barriers to the nascent carbon market. Although they are not exhaustive, the actions proposed here will stand a high probability of establishing a vibrant carbon project sector with inevitable benefits to development and offering a conducive environment.

Create a facilitative, legislative, and regulatory environment. Carbon project development can interact with many sectors of the economy and hence the regulatory environment. By examining the legislative system, regulations that complicate or hinder project formation, approval, and registration can be identified and altered where appropriate.

Successful carbon credit formation, issuance, and trading require credit sellers to explicitly demonstrate their rights to carbon credit sale. Buyers will also want assurance that the right of the sellers will be maintained for the life of that credit issuance. Many countries have varying land and resource tenure laws. For example in some areas, a person may have land-use rights but not mining rights or the right to sell the land. Land tenure issues have been one of the stopping points for many projects under development worldwide. Changing land tenure systems within a country is a large task that may take time to implement; however, regulations can be put in place that allow project developers to more easily obtain land tenure and carbon rights.

Regulations can also be altered with the goal to streamline the country-approval process. In the past, project costs have been unnecessarily raised due to arduous and bureaucratic approval steps. Accordingly, this process could be furthered through a guide to investing in the carbon market, which could describe the approval process for any proposal and the specific criteria under which it will be considered and evaluated.

Source: Walker, Pearson, Munishir, *et al.* (2008)

development of the optimum role for the forestry sector to deal with climate change are predominantly policy based rather than technical.

POSSIBLE ISSUES OF CONTENTION: ADDRESSING PROBLEM OF NON-COMPLIANCE IN KYOTO PROTOCOL

In the case of non-compliance with the binding emission targets set in the Kyoto Protocol, the penalty seems too weak compared to domestic law. In Article 18, one of the most concrete statements is that "failure to meet the quantified commitments in the first period automatically disqualifies a country from participating in the mechanisms and will be penalized by deductions from allowed emissions in subsequent rounds with a 30% penalty factor". Nevertheless, the compliance section was one of the most highly contested issues in the Marrakesh Accords. However, the fact that commitments are legally binding as a principle and the establishment of an enforcement branch in the compliance committee make the compliance package considerably stronger than in most treaties (Grubb 2003). Thus, the Kyoto Protocol appears to have a stringent provision against derivations, as it provides a comprehensive framework that aims at curbing the adverse impacts of climate change.

The United Nations Framework Convention on Climate Change (UNFCCC) and its protocol are international treaties and the parties involved have to implement them accordingly. Following the obligation of the parties, Article 12 of the Kyoto Protocol provides a platform for the developed and developing countries that achieve the quantitative emission reduction targets to collaborate under the CDM. Since the CDM has dual objectives, project participants should comply with rules and regulations (legal framework available at both international and national levels in relation to the implementation of afforestation and reforestation CDM projects) to meet the objectives of sustainable development and other national interests (CIFOR 2005a).

INTERNATIONAL REGULATIONS

Legal instruments to implement carbon projects in the forestry sector include the Kyoto Protocol and its well-known Article 12 on CDM, along with numerous decisions outlining modalities and procedures to implement afforestation and reforestation projects under the CDM. International legal perspectives followed for implementing project activities include the provisions of the UNFCCC, Kyoto Protocol (namely CDM), and Conference of the Parties (CoP) decisions.

Article 12 of the Kyoto Protocol states that the reduction under CDM should be real, measureable, and verifiable and, in the case of forestry projects, that is, afforestation and reforestation projects, it should include sustainable development with maximum transparency and accountability. All these issues have legal implications. But there is no adequate definition of "sustainable development", except vaguely it can be equated to supplying some sort of social benefits to local communities adding to the longevity of the project. As compared with any other sectoral projects, forestry projects require far more complex methodology to calculate and estimate baselines; to develop rigorous methods for measuring, monitoring, and verifying carbon benefits; and to determine accurate methods of measuring socio-economic and environmental impacts in a very cost-effective manner (Davis 2000). Forestry projects, involving different parties such as project developer, consultant, certification, and accepting authority, could be quite contentious, which could be fatal to the project if not foreseen well. Hence, any project proponent would have to be very prudent on this count and should prevent occurrence of any legal complications at the very first place, making provisions in the proposal to face any legal contention in a very amicable manner. These prevailing legal conflicts between different parties may sometimes need to be innovatively resolved by using existing legal procedures.

REQUIRED LEGISLATIONS

Significant attention is required to adequately address the weaknesses of the current legal framework that pose real challenges to forestry carbon projects (Asare 2010), which includes the following issues:

- Poor forest stewardship and governance
- Perverse policy incentives
- Weakly implemented and conflicting laws
- A backlog of land disputes

To explain how the legal and policy frameworks affect forestry carbon projects, Asare (2010) stressed on the need to address how carbon should be regulated and defined, and how rights should be issued. It was speculated that the regulation and management of carbon should rest with the forestry agency; this is because carbon is found in forest biomass and is linked to trees. However, a number of concerns were associated with this choice. The term carbon can refer to "carbon sequestration", "carbon sink", "carbon sequestration potential", and "carbon credits"; hence, "carbon" requires to be clearly defined. A policy decision is needed to classify it either as a security or as a commodity. After carbon is properly defined, the next step is to allocate "rights" (either ownership rights or the economic right to benefit, or both) of the carbon. Who or what is actually driving deforestation and forest degradation is a very important issue for any activity to be permanent.

Carbon rights and ownership issues must be resolved to provide individuals and communities with real and tangible incentives through benefit-sharing mechanisms. The issue of how rural farmers and resource users will be compensated for their efforts to sequester or maintain carbon stocks is more important than that of who owns the carbon. Some examples of effective benefit-sharing mechanism based on sound evidence and experience include the following:

- Distribution of stumpage fees
- Commercial plantation agreements
- Traditional systems of sharing benefits from land

Second, clarity regarding the ownership of planted trees is very important, as larger issues of carbon rights are decided based on this. Therefore, a system of incentives that can help the nation in transforming its perspective and practices in relation to forests and trees may include the following:

- Implementation problems and contradictions in the existing laws need to be addressed.
- Improved forest governance and compliance are critical for both the effectiveness and the equity of market-based or fund-based forestry carbon projects.
- The government could benefit from forestry carbon projects through a taxation mechanism, as opposed to outright carbon ownership; hence, this would require a careful review and design so that incentives for such projects are not reduced.
- The main priority should be on tree tenure reform, so that farmers have positive incentives as regards maintaining timber or other trees as shade in the farming landscape.
- Land and trees that become part of forestry carbon project activities should be surveyed and documented.

SOME PREVENTIVE AND PRE-EMPTIVE STEPS: ANTICIPATING REGULATORY BARRIERS

In the case of some regulatory aspects constraining project implementation, the following anticipating steps may be taken by both parties involved in developing the project design document (PDD) (CIFOR 2005a):

- **Streamlining:** If required, to satisfy the regulatory requirement, the PDD may be streamlined. For example, the area designated for a large-scale afforestation and reforestation CDM project has to be reduced, if it surpasses the requirement. If a project deviates from its original design of being a small-scale afforestation and reforestation CDM and combines several small areas, defunding procedures have to be implemented.
- **Harmonization:** Harmonization is needed if there are conflicting cross-sectoral regulations, in which case a joint letter of understanding may be issued. One should identify or achieve additional or collateral benefits from the project to compensate for any compromise actions involved.
- **Amendment:** Upcoming regulatory instruments (such as regulations related to genetically modified trees) may be "warned" well in advance because afforestation and reforestation CDM project activities do not allow these to be used.
- **Contractual arrangements:** Contracts may be arranged between two governments, between the government and a private entity, or between two private entities when sellers are in the developing countries and buyers in the developed countries (CIFOR 2005a). Two forms of agreements may be followed, each requiring a different legal document:
    - A (government or private) buyer from a developed country agrees with a (government or private) seller from a developing country on the purchase of carbon credits produced by projects implemented in the developing country, with minimum involvement of the buyer. In this case, the buyer only guarantees the market.
    - A buyer is heavily involved in project development, implementation, monitoring, pricing, and issuance of the certificate.

OTHER SOURCES OF DISPUTES

Due to the variable nature of the existing carbon markets and demand for carbon credits from forestry carbon projects, the risk associated with the transactions increases and consequently the probability of disputes between parties engaged in the project could be high. Some of the most common instances of dispute are as follows (Searle 2010):

- Certified emission reduction (CER) delivery failure or shortfall.
- Project failure or delay.
- Delay or failure to obtain regulatory approvals.
- Change of regulation, law, and policies (for example, regarding forest definition).
- Political risks of delay or failure to obtain host country approval.
- Credit and solvency risks and market volatility.

These are some issues related to conflicts and disputes. Forestry or afforestation and reforestation projects are long-duration projects requiring frequent verification at short and regular intervals. Long commitment involves various types of uncertainties, complexities, and also contentious issues such as acquisition of land and dealing with multi-cultural stakeholders with diverse background and diverse incentives. The benefit-sharing mechanism and institutional framework involve critical issues of the project feasibility such as additionality, leakage, and permanence. These can lead to many situations that create a deadlock for the project. These issues can generate problems and legal complexities. Some of the major problematic areas are as follows (Manguiat, Maria, Verheyen, *et al.* 2005):

- Additionality
- Restriction to afforestation and reforestation
- Quantitative restriction of carbon credits
- General principles
- Appropriate legal and institutional system to approve CDM afforestation and reforestation projects
- Land rights issues, title to carbon, and the nature of CERs
- Social and environmental impacts—requirement to assess
- Social and environmental impacts—substantive standards
- Small-scale afforestation and reforestation activities and involvement of local communities

Most of these issues are expected to emerge at the initial stage of project design. These issues, if not given due importance, may cause project complexity or even failure. Hence, to save the stakes of parties involved in the project, the project developer should be ready to resolve legal issues; for this, a project proponent should have clarity on the following few questions before initiating the project (adapted from Manguiat, Maria, Verheyen, *et al.* 2005):

- What is the legal status of the land? Are there any particular restrictions that its legal status imposes on the project site?
- Who owned the land? Is this ownership recognized by law and evidenced by a legal document?
- Who are the bona fide representatives of those who have a stake in the project site? Who should the project investor negotiate with buyers for the CERs?
- Assessing whether the occupants and/or users of the land are different from those who hold legal title to the land. Assessing the rights of the occupants and users of adjacent lands over the potential project site.
- Does the law require the present uses of the land to be incorporated into the project? If any land-use change is involved, does permission need to be sought for such changes, and will the land need to be reclassified?
- What other permits and clearances need to be obtained for the project, apart from the certification that the project assists in achieving sustainable development?

As one proceeds with the project, various other problems are encountered in both implementation and documentation. Hence, a good understanding of legal aspects forms a significant part of forestry projects, in both large- and small-scale projects where these issues can cause great damage at a later stage and can even affect the projected profit, consequently jeopardizing the whole project and interest of the financers.

MANAGING WITH THESE LEGAL COMPLICATIONS

Legal complications tend to happen in a forestry carbon project due to the diversity of stakeholders and its complicated benefit-sharing mechanism. Hence, a project developer dealing with such projects should be ready to take steps to reduce these legal complications. These steps can be taken in two stages, with respect to evolution of such issues: preventive steps and resolution steps.

Preventive Steps

Preventive steps are taken before the occurrence of complications and legal issues; they assist in avoiding evolution of legal issues and resultant complications. Such measures include the following.

Appointment of a Legal Advisor

Forestry carbon projects are full of complexities and procedures; hence, large projects may require a legal consultant to secure a safe path to a forestry project and associated phases by solving various complex issues encountered. The proceedings of forestry carbon projects are technical in nature, and thus advice from legal advisors can really prove fruitful in the resolution of issues as and when they occur. Large-scale forestry carbon projects are more complex than the small-scale projects, and thus may require a permanent legal consultant. On the other hand, temporal appointment can be done in small-scale projects or this job can be outsourced from other legal entities or organizations working in the sector. However, the legal advisor should possess adequate technical expertise and experience to thoroughly perform the functions assigned. Some functions of a legal advisor are as follows (Davis 2000):

- Control of transaction costs
- Protection of the interest of parties involved
- Compliance with international legal and financial standards
- Development of a clear legal definition of "carbon offsets"
- Get oneself clear with international CDM rules
- Learn about the national laws

Therefore, the most important objective of the hired legal advisor could be to assist in designing the project and its related documents in a way that anticipates prospective international and domestic regulatory regimes, along with the insurance of compliance by the parties and the stakeholders involved. Clarity and awareness about associated legal procedures can save both time and effort, resulting into a more successful, well-designed project (Davis 2000).

Having a Well-designed ERPA

Another preventive action that may assist in avoiding the anticipated legal complications is an agreement popularly known as the emissions reduction purchase agreement (ERPA). This is an obligatory agreement between the parties, including the buyer and the seller. This agreement defines all the obligations, rules, and liabilities during the course of transactions. This is done by initially screening/acquiring information by the buyer about the eligibility of the project, estimation of the proposed carbon sequestration by the project, and estimation of the profitability and positive perception among the communities.

This is the document between the involved parties to clarify the interests, obligations, and time frame of negotiations. It includes the basic terms of the transaction but is not a binding commitment after a letter of intent (LoI) is signed. The LoI is usually valid for a shorter time period and terminates when the ERPA is signed. Subsequently, the purchase agreement built on the LoI is structured as a contract, which has to be signed by all involved parties. The ERPA is a binding commitment and includes obligations and liabilities.

An emission reduction transaction (both LoI and ERPA) addresses a list of key parameters that are required to define the object of the transaction, its applicability and validity conditions. From a practical point of view, about the volume specifications, time frame of credit delivery, payment schedule, price settings, and liabilities are explained through some key questions (Neeff, Lisa, Imme, *et al.* 2007).

Resolution Steps: Different Ways

Resolution steps are necessary when conflicts and disputes occur and cannot be resolved amicably, where both the involving parties become inconclusive and thus may require a third party to solve their issues and conflicts. Some of the conflict-resolution mechanisms are as follows:

Arbitration

Arbitration is the dispute-resolution mechanism of choice for carbon credit transactions, where parties involved agreed to do so. Arbitration offers several advantages such as neutrality of the venue. Parties can choose a neutral venue for resolving their disputes rather than being engaged in one party's home court. Arbitrations are generally confidential, cost-effective, and flexible in their procedures of conduction (Manguiat, Maria, Verheyen, *et al.* 2005).

One of the main advantages of arbitration is that the New York Convention[2] provides for the recognition and enforcement of arbitral

[2] The Convention on the Recognition and Enforcement of Foreign Arbitral Awards, also known as the New York Convention, was adopted by a United Nations diplomatic conference on 10 June 1958 and

awards between the countries that are parties to it. These include China, India, Brazil, and Mexico, which together account for approximately 74% of registered CDM projects and 80% of expected CERs (Searle 2010).

Hence, it is always advisable to introduce arbitration clause in any agreement. However, when designing arbitration clause, parties should remember the following points:

- It is important for parties to select a seat or legal place of the arbitration; as a general rule, this should be in a jurisdiction where there is a developed body of arbitration law and limited scope for interference from state courts.
- The parties should select arbitration rules[3] that are widely used in international trade and commercial transactions.
- The parties can also agree for the appointment of an arbitrator or tribunal who has experience in carbon credit transactions and has practised arbitration in the same field.
- In the arbitration clause, parties should also specify a governing law with an objective to save time and cost.
- In the projects where the cost of arbitration can be more than required, a single arbitrator should be preferred against the conventional three arbitrators.

Thus, if arbitration clauses are drafted wisely, they can provide additional flexibility to the procedure adopted in the case of any dispute. While many of the legal issues revolve around forestry law, land use, land tenure, and environmental and contractual laws and policies, other areas of the law can also be brought into the picture. Learning from the past experiences in the context of both national and international development can result into a better-designed project with more efficient handling of legal disputes (Manguiat, Maria, Verheyen, *et al.* 2005). Better-designed documentation and planned legalities will ensure that the project becomes viable and subsequently helps deliver the intended benefits to all the associated stakeholders and communities.

---

entered into force on 7 June 1959. The Convention requires courts of contracting states to give effect to private agreements to arbitrate and to recognize and enforce arbitration awards made in other contracting states. Details are available at <www.uncitral.org/uncitral/en/uncitral_texts/arbitration/NYConvention.html>.

[3] The suggested arbitration rules can be the International Chamber of Commerce Arbitration Rules, the Permanent Court of Arbitration, Optional Rules for the arbitration of Disputes Relating to Natural Resources and/or the Environment.

Mediation

Mediation, a form of alternative dispute resolution or appropriate dispute resolution, is a way of resolving disputes existing between two parties of the same forestry carbon projects. Here, a third party structures the meetings and helps the disputing parties come to a final decision/ conclusion based on the facts about their stand on the projects presented in the discussions. Sometimes, a clause regarding a potential mediator, in the case of any disputes, may be included in the agreement. Mediation is not legally binding; hence, if any one party is not satisfied or feel disadvantaged, it can take the issue to the court, depending on the mediation agreement (Manguiat, Maria, Verheyen, *et al.* 2005).

Arbitration/adjudication (Through Tribunal)

In general sense, a tribunal is any person or institution that has the authority to judge, adjudicate on, or determine claims or disputes, irrespective of whether it is called a tribunal in its title or not.[4] Many governmental bodies are titled "tribunals" to emphasize that they are not courts of normal jurisdiction. Agreement designed in the process of forestry project development or carbon transaction may encompass a clause specifying the "tribunal" in the case of conflicts (Manguiat, Maria, Verheyen, *et al.* 2005).

Court

A court is a form of tribunal, often a governmental institution, that has the authority to adjudicate legal disputes between parties and carry out the administration of justice in civil, criminal, and administrative matters following the rule of law. Courts dispute resolution in both common law and civil law legal systems. All persons are allowed to bring their claims before a court. Similarly, those accused of a crime has the right to present a defence before a court. Conflict resolution through court is less preferred because it directly involves the government, is a cumbersome procedure, and justice is delayed (Manguiat, Maria, Verheyen, *et al.* 2005).

[4] Details available at <www.conservation.org>

11

FORESTRY CARBON PROJECTS AND BIODIVERSITY CONSERVATION

C P Kala

The huge variety and variability of species, including humans, are dependent on each other, and the extinction of any one of these species can directly or indirectly affect the survival of other species. Biological diversity, therefore, is undoubtedly one of the most important components of the ecosystem, which provides countless services in the form of not only food and medicine but also regulating the number of essential services for survival. At present, the human population and subsequent anthropogenic activities are causing havoc to the existence of biodiversity. The causes of biodiversity loss may vary from local to global scales. The localized loss of biodiversity includes over-exploitation and invasion of weeds, whereas the effect of climate change on biodiversity is a globally occurring phenomenon. Realizing the significance of biological diversity, this chapter aims to highlight the vital role of biodiversity and conservation in the long-term sustainability of climate change projects.

IMPACTS OF CLIMATE CHANGE ON BIODIVERSITY

Worldwide, the change in climate is influencing the biodiversity. However, there are arguments and counterarguments on the effects of climate change on biodiversity. One school of thought argues that climate change is of little importance to forests and wildlife because land-use changes and intrinsic properties of biodiversity and nature are mostly responsible for local changes in biotic communities (Lomborg 2001).

This observation seems to ignore the potential complexity of natural systems and biotic changes. Change and extinction of even an individual species may have profound effect on other species as evident from the extinction of Dodo, the flightless bird. The short-term local changes may become important in the longer term if the causal factor continues. The small but constant forces are important for they can alter species interactions and de-stabilize communities (Parmesan and Yohe 2003).

Evidences indicate that changes in plant phenology, range and distribution of species, composition and interactions within communities and even changes in structure and dynamics of ecosystems have been witnessed due to climate change (Walter, Post, Convey, *et al.* 2002). The changes in the timing of seasonal activities of species, for example, earlier breeding, earlier arrival of migratory birds, earlier appearance of butterflies, earlier flowering of plants, and earlier spawning in amphibians, because of climate change have been noticed (Graves and Reavey 1996; Mawdsley, O'Malley, and Ojima 2009; Walter, Post, Convey, *et al.* 2002). Species may fail to avail suitable habitats and food for survival due to local changes in climatic conditions and also the different patterns in associated species that may not be synchronous, resulting in alteration of ecosystem's properties.

The increasing temperature and subsequent changes in resource availability may shift species towards the poles in latitude and upwards in elevation. There are studies showing a typical pattern in species distribution, which includes northward range shifts during the two twentieth-century warming periods (1930–45 and 1975–99), and southward shifts during the intervening cooling period (1950–70; Parmesan and Yohe 2003). In an experiment, climate-driven predicted changes in plants phenology were observed in 85% of the selected species (Parmesan and Yohe 2003).

In the Himalayas, the oak forests are being invaded by the pine forests due to an increase in temperature. The associated species of pine forests may cross frontiers and become new elements of the oak forests. There are phenological changes in terms of early flowering in rhododendrons—one of the major companion species of oak forests in the Himalayas. The treeline zone in the Himalayas is known for its high sensitivity to climate change, and increasing temperature may have profound influence on the flora and fauna diversity of this important ecosystem.

It is an established fact that increasing CO_2 concentration increases plant growth; however, photosynthesis is a very fine-tuned process and depends on the availability of nutrients in the soil and the temperature of the environment. The high altitude areas of the Himalayas are generally nutrient poor, and therefore the rate of photosynthesis in such areas may be negatively affected due to climate change. Many of the Himalayan plant species are used for preparing medicines, and changes in such plant properties, which include change in secondary metabolites, due to change in climate may severely affect the industries dealing with these plant produce (Gairola, Shariff, Bhatt, *et al.* 2010). Change in vegetation distribution may not guarantee a shift in the companion plant and animal species. Especially, the less mobile fauna are assumed to be more affected due to change in the climate.

The sensitivity and capacity of adaptation of ecosystems and biodiversity to climate change may differ across the biome, and thus the effect of climate change on biodiversity may vary accordingly. Although there are multiple factors that influence the ecosystem and biodiversity, climate change is considered a prime factor that changes the biodiversity in an Arctic ecosystem (Sala, Chapin III, Armesto, *et al.* 2000). The biodiversity of cold climatic areas such as alpine and subalpine regions of the Himalayas is, therefore, a matter of concern in the present era of global climate change. Apart from forests and wildlife, the agro-biodiversity is being influenced by climate change. The modern agriculture system, even though it has increased food production, has damaged the natural environment and biodiversity due to increasing use of fertilizers coupled with climate change (Kala 2010a).

MITIGATION STRATEGIES OF CLIMATE CHANGE EFFECTS ON BIODIVERSITY

Species need special care to safeguard ecological integrity. The species facing the risk of extinction need immediate conservation efforts both ex situ and in situ. Apart from habitats, the movement corridors of wildlife species need to be protected. Despite several reports and studies, it is a well-known fact that the data on effects of climate change, especially with respect to biodiversity, are limited. There are also arguments on the suitability of species-oriented conservation planning and ecosystem or biome-based conservation approach. Moreover, the effects of climate change are beyond the boundaries of protected areas such as national parks, sanctuaries, and biosphere reserves. This dilemma is complex, and given the urgency, it is not advisable to wait for solid data set to adequately validate and calibrate models and concepts for protection of nature and biodiversity.

In pursuit of the major objective of the United Nations Framework Convention on Climate Change, Article 2 of the Kyoto Protocol specifically deals with protection and enhancement of sinks and reservoirs of greenhouse gases (GHGs), which ultimately encourage sustainable forest management practices, including afforestation and reforestation. For stabilizing the GHG emissions and hence the effects of climate change on biodiversity, the current state of knowledge on biodiversity needs to be evaluated to develop effective policies, regulations, and laws for mitigation of climate change effects.

PRESUMPTIVE IMPACTS OF CDM ON BIODIVERSITY

The Clean Development Mechanism (CDM) is one of the key mechanisms to address climate change mitigation. The essence of CDM, which is an arrangement under Article 12 of the Kyoto Protocol, is to allow the developed-country parties to jointly undertake emission reduction project activities in the developing countries to meet their GHG reduction commitment. In general, the CDM is an incentive for projects that reduce GHG emissions in the developing-country parties, at the same time generating substantial revenue for the developers of CDM projects. The CDM largely assists the developed-country parties to meet their emission reduction commitments to the Kyoto Protocol and prevent further climate change, apart from assisting the developing countries to achieve sustainable development. The CDM projects may contribute in the sustainable development of countries through the transfer of technology and knowledge, improving social and human capital by creating employment and enhancing living standards, and improving environmental assets by preserving biodiversity and improving local air quality. Despite several positive aspects, CDM projects may also have broader social, ecological, economic, and environmental issues, which need to be tackled wisely.

All the member countries of the Kyoto Protocol are expected to finalize the modalities for afforestation and reforestation projects under the CDM until 2012. Investing in sinks today as per CDM may not help us continue and make high emission cuts tomorrow. Increasing sinks through afforestation and reforestation instead of curbing GHG emissions will add more carbon to the active carbon pool. Looking into the commercial aspects of the CDM for carbon price, being a funding agency, the developed-country party may dictate the developing countries for genetically engineered, monoculture fast-growing species (Sawyer, Meinshausen, and Guilbeault 2003). This will not only negatively affect the native biodiversity but also put pressure on the indigenous agroforestry system. Although plantation and afforestation helps in the reduction of GHG emissions, any human-made plantation cannot create natural habitats. There are a number of intra- and inter-species linkages, which cannot be created by executing monoculture plantation projects. Further, the afforestation CDM projects may remove the legal and customary land and resource use rights of local communities or indigenous peoples over the forestland.

There are ample examples to reveal that the advent of commerce and market forces in plant and animal species generally imposes pressure on the existing biodiversity. The traditional knowledge on the use of biodiversity once brought into market has generally imposed the negative impacts on the population of such useful flora and fauna (Kala 2010b). Although there were traditional norms for the conservation of useful biodiversity, the market forces have superseded the traditional norms of conservation (Kala 2010c). The presumptive market in plantation projects, especially with respect to CDM projects, is other way round, which is an alarm call for the existence of native biodiversity if concerted efforts are not made for its conservation. Plantation not only alters the soil microbes but also changes the environment for the native species. The monoculture plantation of eucalyptus in Central India has influenced not only the groundwater table but also the native biodiversity. It is felt that any CDM project should be planned in view of wise use and conservation of native biodiversity.

BIODIVERSITY CONSERVATION AND CLIMATE CHANGE MITIGATION PROJECTS

Joint implementation, carbon trading, and CDM of the Kyoto protocol encourage countries to design and implement projects related to climate change mitigation and sustainable development. Realizing the importance of biodiversity in any natural ecosystem, as discussed earlier, it is expected from any such project to emphasize the need of biodiversity conservation.

To meet this goal, the entire evaluation system of such projects should be based on some quality standards focusing biodiversity conservation issues. Plantation of native tree species should be encouraged by giving more values to such projects. Rather than focusing on single-strata tree plantation, multi-strata plantation may be encouraged for providing space to different species to flourish. The guidelines for sanctioning climate change projects must include and elaborate strategies to mitigate pressure of such projects on the diversity, richness, and habitats of native species. The existing standard, for instance Climate, Community, and Biodiversity (CCB) Standards (CCBA 2008), can be followed, which stresses to incorporate the status and conservation of important biodiversity parameters (such as critical habitats, threatened species, high conservation values, endemism, and breeding areas) as important components of the project. As suggested in the CCB Standards, any such project should also have a comprehensive plan for documenting the changes in biodiversity resulting from the project activities. In general, climate change mitigation projects should be designed in a way to provide synergy between sustainable development and biodiversity conservation.

12

ADAPTING TO CLIMATE CHANGE: ROLE OF FORESTS AND THE FORESTRY SECTOR

U K Sarma

IMPERATIVE FOR ADAPTATION STRATEGIES IN THE FORESTRY SECTOR

Much of the discussions and efforts on climate change and forestry sector have been through the lens of mitigation. Little attention has been devoted to the need for adaption of forest ecosystems and the forestry sector to climate change, and the adaptation of people depending on the forest ecosystem. This needs to be given due priority considering the fact that forests are highly vulnerable to the adverse impacts of climate change, which in turn could have significant repercussions on the livelihoods of forest-dependent communities. This assumes greater importance in a developing country such as India, which has a large population depending on forests for livelihoods. Forests also need to figure prominently in climate change adaptation policies because they have a key function in reducing the vulnerability of society to losses caused by climate change. At the local level, forests provide ecosystem services and will be essential in maintaining people's environments and livelihoods as the climate changes. At the same time, forests too need help to adapt to the changing climate. Tropical forests in some parts of the world are already affected, most obviously by changes in temperature and rainfall that are leading to a greater risk of fire. Although the scale of natural disasters, predicted to increase with climate change, depends on an aggregation of factors, in many cases, ecosystem preservation and forest protection can greatly reduce their impacts. Coastal mangroves, floodplains, and forests, for instance, offer physical defence by buffering land, communities, and infrastructure against natural hazards (Mansourian, Belokurov, and Stephenson 2009).

Adaptation to climate change encompasses changes made in natural or human systems to reduce the damage caused by climate change or to exploit the benefits (Guariguata, Cornelius, Locatelli, *et al.* 2010). At the same time, it needs to be kept in mind that adaptation represents an enormous challenge in the forestry sector due to two reasons—the first is the lack of knowledge about the sector's vulnerability and the second is the lack of experience in the field (Robledo and Forner 2005).

BASIC CONCEPTS: IMPACTS, VULNERABILITY, AND ADAPTATION IN THE CONTEXT OF FORESTRY

A discussion on adaptation in the context of forests and the forestry sector would benefit greatly from the elucidation of a few basic concepts central to the issues at hand: impact, vulnerability, and adaptation.

Impact is defined as the consequences of climate change on natural and human systems, which could be either detrimental or beneficial (Levina and Tirpak 2006). Owing to the multiplicity of geographical, environmental, social, and economic conditions, an effect may bring positive consequences to one social group and negative consequences to another (Robledo and Forner 2005). The Intergovernmental Panel on Climate Change (IPCC) *Third Assessment Report* (IPCC 2001b) concludes that forest ecosystems will be affected by climate change in a number of ways: greater risk to populations of many threatened species; alteration in species composition and dominance; enhanced global timber supply; decreased biomass productivity in dry forests, woodlands, and rangelands; cases of forests or heathlands replacing some wetlands; disruption of those forests overlying permafrost owing to thawing; and higher incidence of extreme weather events (such as droughts coupled with warming, leading to increased occurrence of fires). The extent to which an ecosystem is likely to be impacted by climate change depends on a host of factors: magnitude of the change, probability of a change occurring and also of its intensity, rate at which change occurs, duration of change, and the characteristics of the system, which refer to the capacity of the system to adapt to the changes.

Adaptation to climate change refers to adjustment in natural or human systems in response to actual or expected climatic stimuli or their effects, which moderates harm or exploits beneficial opportunities (IPCC 2001a). Adaptive capacity indicates the degree to which a system has the capacity to generate such adjustments. Adaptation measures are concrete and specific actions that aim at the reduction of a system's vulnerability or improvement in its adaptive capacity. According to Smit and Skinner (2002), depending on the nature of the timing, adaptation measures could be anticipatory (implemented before the impact has occurred), incidental (implemented while the impact is occurring), and reactive (implemented after the impact has occurred or started to occur). On the basis of duration, it could be tactical (short-term application with results occurring in five years) or strategic (implemented in the medium and long term) and, on the basis of who has the responsibilities, public or private. According to scale, adaptation measures could be localized and disseminated. However, adaptation to climate impacts in the case of forests is rendered difficult due to a number of factors such as the fact that forest ecosystems require the longest response time to adapt, say through migration and regrowth (Leemans and Eickhout 2004). The IPCC *Third Assessment Report* (IPCC 2001b) recognizes that owing to poorly developed institutions, markets, and technology transfer pathways and the lack of financial resources, the developing countries, particularly forest-dependent communities, have low capacity to cope with or adapt to adverse impacts.

Vulnerability is another related concept that refers to the degree to which a system is susceptible to, or unable to cope with, adverse effects of climate change, including climate variability and extremes. Vulnerability is a function of the character, magnitude, and rate of climate variation to which a system is exposed, its sensitivity, and its adaptive capacity (IPCC 2001b). The greater the adaptive capacity of a system (autonomous adaptation), the lesser is its vulnerability. Human beings have varying capacity to foresee impacts of climate change and to generate responses aimed at minimizing them. Planned adaptation, which supplements autonomous adaptation, depends on various factors: economy and equity, technological development, human resources, and infrastructure. An adaptation process starts with a vulnerability assessment, which connects the expected impacts of climate change to the social, environmental, and economic realities of a region, thus leading to identification of the needs and priorities for action (Robledo and

Forner 2005). The United Nations Development Programme specifically talks about socio-economic vulnerability as an aggregate measure of human welfare, which integrates environmental, social, economic, and political exposure to a range of harmful perturbations (cited in Levina and Tirpak 2006). Forest dwellers are the most vulnerable as they have the least resources and the least capacity to adapt.

IMPACT, VULNERABILITY, AND ADAPTIVE CAPACITY OF INDIAN FORESTS AND FORESTRY SECTOR

Climate projections for India indicate that nearly 68%–77% of the forested grids are likely to experience certain changes; for example, loss of area under a given forest type and replacement by another from the prevailing forest type by 2085 (Ravindranath, Joshi, Sukumar, *et al.* 2006). There may also be a shift to wetter forest types in the north-eastern regions and drier forest types in the north-western regions of the country. All these will likely have a significant impact on wildlife species, net primary productivity, growing stock (biomass), and regeneration. According to Ravindranath, Joshi, Sukumar, *et al.* (2006), the anticipated impacts of climate change might affect about 70% of the vegetation in India, which may adapt less optimally to its existing location, thereby being more vulnerable to the adverse climatic conditions as well as to the biotic stresses to which it may be exposed from time to time. Forest-dependent communities in India are going to find themselves very vulnerable to the adverse impacts, whose livelihoods are already bearing the brunt of forest degradation and loss. Thus, there is also an urgent need to deal with the problems of forest fragmentation (leading to loss of biodiversity and hampering migration of species); forest degradation; non-sustainable extraction of timber, fuelwood, and non-timber forest products; dominance of monoculture species under afforestation; or the absence of fire protection and management practices (Ravindranath, Joshi, Sukumar, *et al.* 2003). However, the Indian forestry sector is still saddled with policies and programmes that are not appropriately geared to needs of the times and thus could end up increasing the vulnerability of forest ecosystems.

The presence of a given type of ecosystem is primarily determined by a combination of temperature and rainfall patterns. A change in climate variables, therefore, implies a change in the structure and functioning of ecosystems. The general impacts of climate change on forest ecosystems include the following (Scholes and Linder 1998):

- Changes in the location of areas suitable for the growth of certain species (shift or disappearance of some productive systems).
- Increases or decreases in the production of timber and non-wood forest products (NWFPs) per unit area.
- Changes in the types and incidence of pests and diseases affecting tree and plant species.
- Altered ecosystem functions (biochemical cycles).
- Increased or decreased nutrient retention.
- Changes in the reproductive cycles of species.
- Changes in the value of a system as a tourist attraction.

Taking into account the likely adverse impacts of climate change on Indian forest ecosystems and the high vulnerability of both the forestry sector and the social groups dependent on them, there is an urgent need to put in place strategies that facilitate adaptation. Modified silvicultural activities in natural forests may be required in the context of climate change. The range of approaches for enhancing adaptive capacity to climate change is perhaps wider for planted forests than for natural forests, as both genetic material and silvicultural treatments can be adjusted more readily (Lamb, Erskine, and Parrotta 2005). Fire management in the context of a changing climate also deserves attention.

Urgent, anticipatory action is also the need of the hour because there is a time lag between climate change and its impacts, and there is a long gestation period between adaptive practices and their impacts to become visible in the forestry sector.

FRAMING ADAPTIVE RESPONSES AND ENHANCING CAPACITY OF FORESTS AND FORESTRY SECTOR

According to Robledo and Forner (2005), adaptive measures in forest ecosystems should be based, to a large extent, on enhancing the sustainable production, treatment, and commercialization of forest goods and services. In their opinion, strategies have to be figured out for adaptation projects to create alternatives for social groups dependent on fuelwood; these strategies could include promotion of small plantations for fuelwood production, improved practices to produce charcoal, and so on. The same holds for NWFPs, on which forest communities in countries such as India are highly dependent. For example, in the rural areas of Madhya Pradesh, India, between 40% and 60% of the annual income comes from commercializing NWFPs (Tewari and Campbell 1996). Agroforestry systems, which integrate food and wood production and supply many environmental services, could be harnessed as an important adaptive measure.

The Convention on Biological Diversity (CBD) has recognized the imperative of an ecosystem approach, which is based on the close relationship among different natural resources and their management. Going by this approach, an adaptation project in the forestry sector cannot afford to miss out on the links with other natural resources and their management. The Report of the Second Ad Hoc Technical Expert Group on Biodiversity and Climate Change (2009), constituted by the CBD, recommends an "ecosystem-based adaptation", which integrates the use of biodiversity and ecosystem services into an overall adaptation strategy. This strategy can be cost effective; can generate social, economic, and cultural co-benefits; and can contribute to the conservation of biodiversity. It provides services that include sustainable management and conservation and restoration of ecosystems to help people adapt to the adverse effects of climate change. Some ecosystem-based adaptation activities are, for example, maintenance and/or restoration of mangroves and other coastal wetlands to reduce coastal flooding and coastal erosion, sustainable management of upland wetlands and floodplains to maintain the flow and quality of water, conservation and restoration of forests to stabilize land slopes and regulate water flows, establishment of diverse agroforestry systems to cope with changes in climatic conditions, and so on.

In the Indian context, considering the current state of knowledge and the uncertainties involved, Ravindranath et al. (2003) recommended adoption of only "no regret" or "win–win" activities and a few "precautionary" adaptation policies and programmes. These include the following:

- Revamping forest policies to incorporate climate concern in the long-term forest policy formulation and forest working plan process.
- Effective implementation of existing polices, legislation, and programmes.
- Linking protected areas, wildlife reserves, and reserve forests.

- Expanding area under joint forest management with community empowerment and capacity building.
- Policy support to afforestation and reforestation and to increase the area under timber and fuelwood to reduce pressure on primary forests.
- Policy support to fuelwood conservation programmes to reduce pressure on forests.
- Following certain silvicultural practices that could reduce vulnerability and enhance resilience such as mixed-species forestry, anticipatory planning of species along latitudinal and altitudinal gradient, in-situ and ex-situ conservation of plant and animal species, implementation of fire prevention and management practices, adoption of short rotation species and practices, sustainable harvest practices for timber and non-timber products.
- Capacity building of research institutions, forest departments, non-governmental organizations, and local communities.

SYNERGY BETWEEN ADAPTATION AND MITIGATION IN FORESTRY SECTOR

There is a need for achieving better synergy between mitigation and adaptation activities in the forestry sector. According to Ravindranath (2007), it is first necessary to ensure that mitigation programmes or projects involving afforestation and reforestation, avoided deforestation, and improved forest and grassland management practices do not increase the vulnerability of forest ecosystems and plantations. At the next level, many mitigation projects could incorporate adaptive practices with many strategies such as forest conservation, protected area management, afforestation and reforestation, offering opportunities for both mitigation and adaptation, and at little or no extra cost. For example, effective management of a protected area arrests forest degradation and facilitates conservation of carbon sink in the forests. At the same time, protected areas help protect biodiversity, which reduces vulnerability to climate change. Linking different protected areas enables migration of flora and fauna, which helps in the adaptation process. Unfortunately, knowledge on many significant aspects, which underscores the need for sustained research on these issues, is lacking.

ENSURING GOOD ADAPTATION PRACTICES IN FORESTRY SECTOR

The effectiveness and success of policy frameworks for adaptation also depend on factors such as decentralization, political eagerness, involvement of civil society, and level of awareness. An adaptation policy framework must be flexible enough to allow the decentralized implementation of adaptation measures. At local levels, socio-economic and environmental variability is smaller and the implementation of adaptation measures and strategies can be both effective and efficient. It also increases local stakeholders' sense of ownership over the resource and their responsibility towards it (provided that local people understand the seriousness of the climate change problems, which otherwise is a task in itself). The management and use of forest ecosystems, as well as the decision-making process, must be based on the knowledge of regional and local priorities. The knowledge and experience of people directly dependent on forest ecosystems must be acknowledged as a valuable element in the design and implementation of adaptation measures. Linked to this is a political willingness, which is the basis for governments to acknowledge the importance of forests within the national context, so that adaptation policies create room for action at the local level. Also, be it at the local or national level (which is the sum of different local environments), capacity building would be a prerequisite to implementation of adaptive measures—anticipatory, incidental, or reactive (tactical or strategic). Capacity building would enable the existing social system, especially its institutions, to implement actions that help adapt to climate change. Finally, the identification and establishment of financing sources must be part of an institutional framework geared towards adaptation, as it entails a cost. An obstacle to financing in adaptation is the uncertainty regarding the impacts of climate change (whether they will happen or not, their intensity, and their costs). Although there is a wide range of financing sources for adaptation-related activities, precautionary measures such as flexibility in objectives, openness to co-financing, willingness to tap the private sector, and clarity regarding risk associated with maintenance cost and covering the same with the project budget would help in providing financial sustainability (Robledo and Forner 2005). However, in some cases, the private sector may decide to invest in adaptation measures purely on the basis of commercial self-interest (for example, modifications to tree improvement programmes).

Last but not the least, there remains the absolute necessity of good governance in the forestry sector, including transparency in the policy framework development and implementation process, which may be crucial to the effective implementation of adaptive measures and making the sector robust, enabling and capable of taking head-on challenges in the wake of climate change.

13

FUTURE OPPORTUNITIES FOR INDIAN FORESTRY SECTOR: THE POST-2012 SCENARIO

H S Gupta

DEVELOPMENTAL IMPERATIVES FOR INDIA: EMERGING ISSUES AND SCENARIOS

Over the years, India has transformed into an economy fuelled by industrialization despite majority of its population being dependent on agriculture. Thus, it is imperative for India to pursue its developmental goals and high economic growth rate, for at least 25 years, to eradicate poverty (Korppoo, Linda, Johannes, *et al.* 2009). India is strenuously trying to increase the production and generation of electricity because access to electricity can reduce the gravity of poverty. Electricity is a critically sought-after resource for development; in spite of recent advances in the energy sector, well over 50% of the rural population in India[1] does not have access to electricity. India is counting on its fast economic growth and thus is not in a position to dispense with industrialization despite an increase in greenhouse gas (GHG) emissions. It accounts for approximately 4.7% of total global emissions, which makes it the fifth largest emitter in the world (Fujiwara and Christian 2010). Incidentally, India is largely dependent on coal for electricity generation, and this consumption of coal makes it one of the biggest emitters of GHG gases. Economic development, infrastructure, electricity, poverty, and employment are some of the biggest concerns of

[1] Details available at <www.powermin.nic.in>

India. Economic development is also seen as the principal way to foster adaptation and mitigation strategies to climate change, since adaptive capacity is dependent on the extent of economic development of the rural and urban poor in India. Political and traditional views suggest that environment, although important, cannot be given priority over other significant developmental issues. The per capita emission in India is still low as compared to other developing countries; thus, the Indian government has maintained a fruitful stand in the fight against climate change in this context. Moreover, there is a perception in the developing countries that climate change is caused by the developed countries and so only they should be pioneering in the fight against global warming and other environmental issues. Subsequently, India in its National Action Plan on Climate Change (NAPCC) has clearly suggested that national development is a priority, with an objective to build up its capacity for coping with the adverse impacts of climate change.

Consequently, India has resolutely refused to make binding commitments to reduce its GHG emissions, or other commitments such as targets for energy intensity or share of renewable energy. Indian delegates have always argued that taking on such mitigation commitments could "keep the developing countries poor for another three generations". The Indian Prime Minister, however, gave a political assurance at the G8 Summit of Heiligendamm that while India's carbon emissions would inevitably rise in the short and medium term, India would ensure that it would allow its per capita emissions to exceed the average of the developed countries' per capita emissions. This pledge was repeated in the NAPCC and India's recent submissions to the United Nations Framework Convention on Climate Change (UNFCCC) (Korppoo, Linda, Johannes, *et al.* 2009).

India firmly believes in "common but differentiated responsibility" of the Kyoto Protocol and is determined to follow the same stance in the future. India has also constantly emphasized that equity relative to emission is the way forward in the climate negotiations. Consequently, India supports the current burden-sharing structure of the UNFCCC and the responsibilities. Accordingly, the Indian National Environmental Policy lists the key elements of India's response to climate change as (1) the principle of common but differentiated responsibilities, (2) reliance on multilateral approaches, (3) equal per capita entitlements to global natural resources, (4) overriding priority of the right to development, and (5) identification of key vulnerabilities of India to climate change (GoI 2006; Korppoo, Linda, Johannes, *et al.* 2009).

India's view is that the "shared vision" of the Bali Action Plan has already been written in the Convention (mainly in Articles 2 and 4), and that the post-2012 negotiations should be revolving around the question of how the implementation of the Convention could be enhanced (Korppoo, Linda, Johannes, *et al.* 2009). Another important aspect for the Indian equity perspective is advocating per capita rights to global environmental resources. The Indian delegation wishes to keep this topic on the table, although a per capita system is politically out of reach and will not be operationalized in the post-2012 agreement (GoI 2006). Lastly, the equity issue that India has been defending vociferously is linking developing-country actions and reporting to northern financing. The argument is based on the so-called "linking clause" of the Convention (Article 4.7). In the thirteenth Conference of the Parties (CoP) in Bali, the Indian delegation played a key role in defending this linkage, which was somewhat reinforced in the Bali Action Plan (Article 1 b ii).[2]

India is still waiting for the delivery of promises made in Rio (Earth Summit), that is technology transformation and funding of mitigation strategies in the developing countries. The CoP-13 in Bali saw the issue of technology transfer gaining heightened attention, as it was moved to the Subsidiary Body on Implementation from the Subsidiary Body on Scientific and Technological Advice, where it had long been housed. In the CoP-14 in Poznan, India once again outlined that Annex I parties were obliged to provide new, additional, adequate, and predictable financing to developing-country parties to implement the proposals for mitigation and adaptation.

So far, in the negotiations for the post-2012 climate agreement, India has been only open to implementing those policy measures that are aligned with its own developmental priorities. The position has two major indications. First, if resources used to combat climate change are framed as money not spent on poverty eradication, it is evident that mitigation activities should be financed by the north. Second, the actions taken by the developing countries should be development actions with climate co-benefits, and not vice versa. Indian negotiators have also presented an interesting variation on the development versus environmental stewardship argument—that is, unless the current government sustains

[2] Details available at <http://unfccc.int>

the highest possible economic growth, future generations will inherit an earth that is highly vulnerable to climate change (Korppoo, Linda, Johannes, *et al.* 2009).

EMERGENCE OF REDD

The option of Reducing Emissions from Deforestation and forest Degradation (REDD) in developing countries not only averts the worst consequences of global warming but also generates enormous co-benefits for biodiversity conservation and sustainable development. A review of the literature and analyses of data show that REDD could substantially decrease the severity of climate change. Similarly, information related to costs reveals that REDD is an inexpensive approach compared with emission reductions in the energy sectors of industrialized countries. The costs per tonne of reducing current carbon dioxide emissions from deforestation by half—even with pessimistic assessments, and including opportunity costs as well as REDD's implementation, transaction, administration, and stabilization costs—are less than a third of the current capped carbon market prices (Cacho 2006). As per the conservative estimates, in 2020, an annual funding of $5 billion could reduce deforestation emissions by over 20%, $20 billion by 50%, and $50 billion by 66%. The latter level of funding is equivalent to only 0.13% contribution of the developed countries' annual gross domestic product to REDD (the commitment made by Norway at the Bali convention in December 2007). An estimate from the cost curves of the US Lieberman–Warner Bill, which determines the potential emission reductions from its market-linked funding, reveals that, in 2020, the bill's allocation of 2.5% of allowance revenues for REDD could reduce emissions by an amount equal to 9% of the United States' total emissions in 1990. However, political and institutional constraints are likely to considerably resist the development of a worldwide REDD system (Doucher 2008).

DEVELOPMENT WITHOUT DEFORESTATION

Reducing emissions from deforestation is a part of the needed pattern of development, in industrialized and tropical forest nations alike, for limiting dangerous climate change as well as for providing the basic necessities of life—including food, shelter, health, and education—to all human beings (Doucher 2008). In other words, the vision is that these two historic transformations can be supported by a path towards development without deforestation. Success will require financing (from the wealthy countries of the world) that offers not only the possibility of cutting a major source of GHG emissions but also the potential to provide a substantial transfer of resources from north to south for sustainable development. It is rightly expected that by 2050 we would have reduced developed-nation emissions by 80%, reduced emissions from tropical deforestation to zero, and ensured that every child born on earth would be able to live a life of security and dignity. Different countries will contribute in diverse ways to these efforts with options for reducing deforestation that include parks and reserves, confirmation and enforcement of indigenous land rights, revision of road-building plans, and cracking down on illegal logging. Even programmes in urban areas (such as improvement in health care, building schools) can alleviate the pressures on forestlands and thus reduce deforestation. Similarly, developed countries can contribute through a variety of mechanisms, including the carbon market, official development assistance, international lending institutions, the proceeds of allowance auctions or carbon taxes, and the allocation of allowances. Although development without deforestation is just one aspect of the struggle to avoid the worst consequences of climate change, it transcends that struggle. REDD is not only a way to confront the challenge of global warming but also a part of a new path towards sustainability for rich and poor nations alike. It represents a new phase in the history of the planet (Doucher 2008).

"REDD" PATH OF CLIMATE CHANGE: THE DIRECTION FOR THE FUTURE

One of the main limitations of the Kyoto Protocol with regard to forestry was that no other aspects of forestry except afforestation and reforestation were included in the Kyoto Protocol although emissions due to deforestation and degradation constitute 17% of the total GHG emissions worldwide (Nabuurs, Masera, Andrasko, *et al.* 2007). Deforestation is one of the most prominent and widespread problems in tropical countries, primarily because of the absence of any incentive mechanism for communities who are engaged in illegal and unsustainable felling to relinquish it. Secondly, the perceived opportunity cost of maintaining the forest is high; people pursue immediate economic benefits accrued due to illegal logging and timber trading rather than the indirect benefits ascertained by maintaining forests. Conservation of forests is also important because, in addition to environmental benefits, they supply some special social co-benefits such as employment and means for sustenance and livelihood. Despite having such a positive impact on both communities and environment, forestry is just being limited to afforestation and reforestation in the Kyoto Protocol's Clean Development Mechanism. These advantages of forest conservation and deforestation have forced environmentalists to argue that REDD could be a low cost and effective strategy to mitigate climate change (Sathaye, Makundi, Dale, *et al.* 2007).

GLOBAL EFFORTS TO SHAPE REDD

The member countries of the Coalition for Rainforest Nations such as Papua New Guinea and Costa Rica raised their efforts to include REDD in the policy framework in the CoP-11 in 2005 (Aggarwal 2010). The objective of the proposal was to force the developed countries to bear the associated opportunity cost to control deforestation and degradation in the developing countries, and they should bear this cost given their historical emissions burden. Therefore, a proposal on REDD was sought to create a mechanism by which the developed countries will fund the reduction in deforestation and degradation of forests in the developing countries. This issue took centre stage at the CoP-13 in Bali in December 2007, with a strong support from the host country (Aggarwal 2010).

The REDD mechanism is expected to bring a new set of actors with fresh resources and finances, as well as opportunities for sustainable forest management. The concept of providing a financial incentive for forest conservation through international financial transfers connected with carbon, or REDD in the developing countries, is now relatively high on the international climate agenda. Still there are many risks and issues, such as risks to governance, rural livelihoods, and the integrity of the Kyoto Protocol, associated with these models.

There have been various submissions from different governmental organizations and non-governmental organizations on their positions on REDD, and related methodological, financial, social, and institutional issues. Discussions on methodological issues include whether REDD projects should be implemented at the national level or sub-national level, how regularly these should be monitored, what should be the baseline year, and so on (Kanninen, Murdiyarso, Seymour, *et al.* 2007).

Likewise, issues related to governance of forests, rights of indigenous communities, and equity among various stakeholders have been raised by researchers and civil society groups. It has been argued that implementation of REDD could affect the rights of indigenous communities over the forests (Rawles 2008). All these issues and prospects related to REDD have been attempted to be analysed in this chapter. First, issues have been categorized into design and non-design types and they have been analysed. Subsequently, prospects such as low-cost option and multiple benefits have been discussed. Issues pertaining to REDD can be categorized into design and non-design issues, which need to be analysed carefully. Design issues pertain to the structure of the mechanism and are largely technical in nature, whereas non-design issues relate to broader social, political, and environmental concerns (Table 13.1).

REDD has the potential of delivering multiple environmental and developmental benefits; however, a number of design and other issues need to be addressed before it is implemented. Macro-level issues are design issues related to policy framework, scale, baseline, monitoring, leakage, and permanence. Despite being complicated, these issues could be addressed through viable technical solutions and political agreements, because they are largely technical and political in nature. On the other hand, non-design issues such as the rights of indigenous communities and ecosystem diversity are quite complex and multilevel, and have social, political, economic, and environmental ramifications. REDD projects will be effective when these issues are carefully addressed both at the

Table 13.1 Design and non-design issues related to REDD

Design issues	Non-design issues
Policy framework	Source of funding
Scale	Right to communities
Sub-national/project-level approach	Carbon-centric forest management
Baseline	Ecosystem diversity
Monitoring	Institutions for implementation and distribution of revenues
Leakage	REDD offsets emission reduction targets of developed countries
Permanence	Convergence of interests

REDD – Reducing Emissions from Deforestation and forest Degradation policy and at the implementation levels. Development and livelihood concerns of indigenous groups and forest-dependent communities as well as respect and recognition for their sociocultural and religious rights should constitute an important part of REDD. Otherwise, it is very likely that powerful interests would take over the resource for carbon revenues and these groups will be further marginalized. As a result, social conflicts will arise, leading to serious implications for forest management. Hence, it is imperative that during the design of inclusive REDD policy and its effective implementation, these issues are addressed.[3]

REDD is unlikely to achieve significant emission reductions from deforestation unless it explicitly includes trees in the forest frontier and agriculture–forest mosaics.[4] Sound understanding of the drivers and trajectories of land-use change and the potential impact of alternative development pathways are required for the development of effective REDD strategies. Coherent multi-sectoral approaches are needed to act on the most important drivers of deforestation and support the development of tree-based farming systems and enterprises. Specific financial incentives are required for high-carbon land-use options, which will help meet successfully carbon sequestration, biodiversity, and poverty alleviation goals (Minang, Swallow, and Meadu 2008).

The author argues that while governments will face trade-offs in deciding the ways to respond to climate change, they should also look for the opportunities to capture synergies between approaches that meet both short- and long-term objectives. Improved forest management provides many such synergies, along with opportunities for jointly advancing countries' adaptation and mitigation objectives. The author also emphasizes that governments are responsible for informing citizens about the adaptation and mitigation choices to be made, and encouraging public participation through establishment of democratic processes. Climate change presents governments with many critical choices, including how to meet the political and moral imperatives of responding to today's urgent needs, and at the same time, to meet the needs of the future by investing sufficient resources (Seymour 2011).

---

[3] Details available at <www.fao.org>

[4] Details available at <www.asb.cgiar.org>

BETTER FOREST MANAGEMENT: A WIN-WIN SOLUTION

Improved forest management will help in reducing the current and future vulnerability to climate change and will also advance adaptation and mitigation objectives jointly. Forests can safeguard communities and societies more broadly from the effects of current climate variability such as droughts, storms, flooding, and landslides, in the short term. For example, forests can help protect coastal areas from storms and waves, and forest-based ecosystem services can help regulate hydrological flows during the years having abnormal rainfall. During drought, which affects agriculture, forest-based food and other products can be consumed or sold for income. In addition, forests also help people adapt to climate change in the longer term (Seymour 2011).

However, when forests are degraded or converted for other uses, the supply of goods and services from them is compromised. Degraded forests are less resilient to climate change compared to intact forests. For example, forest areas that have experienced fragmentation or unsustainable logging practices are more vulnerable to fire. Accordingly, investment is required in the sustainable management of forests to prepare for next year's drought as well as to anticipate the 2050 shift in rainfall patterns (Seymour 2011).

BENEFITS OF ECOSYSTEM-BASED APPROACHES

Many approaches to climate change adaptation involve hard infrastructures such as building dams and reservoirs to control floods and to store water. These approaches can be replaced with cost-effective ecosystem-based approaches to adaptation, which include maintenance of forest cover to conserve natural hydrological functions. Such approaches add more flexibility to adaptation strategies. In light of the many current benefits provided by forests, such ecosystem-based approaches to adaptation are likely to belong to the "no regrets" category (Seymour 2011).

Investment in sustainable forest management as an adaptation strategy yields a "double dividend" related to mitigation; carbon sequestration and storage are some recently appreciated services provided by forests. Deforestation and forest degradation account for up to one-fifth of the current global carbon emissions. Hence, preservation of forests must be included in the climate change mitigation strategies at the national and global levels (Seymour 2011).

POTENTIAL TRADE-OFFS: WINNERS AND LOSERS

However, forest ecosystems are in danger as a result of climate change and exploitation pressures to meet the current food, fuel, and fibre needs. Protection of forests should be seriously considered at the local and global levels; however, this may pose difficult choices between current and future welfare, and create "winners" and "losers" across different geographic locations.

At the level of particular landscapes, the ecological integrity (and thus resilience) of forests needs to be protected from threats posed by conversion to other uses, fragmentation, and degradation due to excessive harvesting of timber, wildlife, fuelwood, and other resources. Protection of forests from deforestation and degradation will reduce emissions. Such protection at the global level will help reduce the risk of catastrophic climate change that could damage the current ability of forests to store and sequester carbon (Seymour 2011).

Financial mechanisms being mobilized under the aegis of REDD+ may provide a source of finance for such protection, and for compensating communities for any loss of current income that such protection entails. Forest protection efforts can either help finance rural development or make some stakeholders worse off easily, depending on the sharing of REDD+ benefits at the national and local levels. Thus, there is an important trade-off between imposing risks on some of the world's most vulnerable communities in the short run and the risk of no action to reduce forest-based emissions, which benefits the global community as a whole in the long run.

At the global level, REDD+ funds are likely to be targeted to the "high-carbon" humid forests of the Amazon Basin, the Congo Basin, and Indonesia. As a result, other sources of finance will be needed to address adaptation challenges in areas most affected by climate variability, including the dry forest areas of Sub-Saharan Africa. To minimize zero-sum trade-offs, governments should build adaptation synergies into the design of forest interventions funded by REDD+, and should reserve scarce adaptation funds for targeting action in "low-carbon" forests (Seymour 2011).

THE GOVERNANCE CHALLENGE

In many countries, poor governance and weak institutions in the arena of forest management present a cross-cutting challenge. The adaptation and mitigation benefits of sustainable forest management can be realized when governments will challenge vested interests in the status quo and will invest in building necessary institutional infrastructure, which is currently lacking at all levels. Such infrastructure should connect local communities with higher levels of government; this will assist in inter-sectoral collaboration and empowering citizens to voice their choice in the design of adaptation and mitigation strategies. But this investment entails few short- or long-term trade-offs, since many of the same governance and institutional capacities are needed to respond to immediate development needs, and disaster risk reduction is similar to those necessary to prepare for climate change in the long run (Seymour 2011).

However, issues such as making trade-offs between current and future welfare, allocating costs and benefits across stakeholder groups, and decisions about risk taking and management are politically decided. Thus, one of the most important "no regrets" investments for all governments to make is to inform their citizens about the adaptation and mitigation choices ahead and establish democratic processes to allow the people to participate in making those choices wisely (Seymour 2011).

CONCLUSION

Currently, most of the research and discussion of low-carbon economy come from industry, particularly the energy industry fields. However, the most promising forestry has been ignored. Regarding the realities of forest, with its special functions, it is one of the most important and easiest solutions to climate change mitigation and low-carbon economy development. It is a solution beneficial to promote sustainable development of national economy, safeguard national eco-security, and pledge human welfare by developing low-carbon economy at low cost, with comprehensive benefit, and real CO_2 reduction. As a result, it is significant to make great efforts to implement major construction of ecology project, to enhance capacity of carbon credit, to promote forest planting and afforestation, to develop forest carbon sequestration, and to reinforce forest carbon functions. It is crucial to stress the significance and functions of the forestry in developing low-carbon economy as well as carry out researches on forest to promote low-carbon economy.

Appendix 1

List of Agencies Facilitating Forestry Carbon Projects

Name	Organization type	Project type	Website
3C Group	Broker	Afforestation and reforestation plantation, energy efficiency, off-grid renewable energy	www.3c-company.com/en
Action Carbone	Retailer	Afforestation and reforestation mix native; methane: coal mines; energy efficiency; renewable energy credits	www.actioncarbone.org/main_fr.php
Bioclimate Research and Development Ltd	Project developer	Afforestation and reforestation mix native	www.bioclimate.net
Carbonfund.org	Retailer, wholesaler/aggregator, broker, project developer	Unspecified mix	www.carbonfund.org/site
The Carbon Neutral Company	Retailer	Afforestation and reforestation plantation; afforestation and reforestation mix native; methane: livestock, coal mines; energy efficiency; off-grid renewable energy; mixed/not specified	www.carbonneutral.com
Cleanairpass	Retailer	Afforestation and reforestation mix native; methane: livestock	www.cleanairpass.com
Climate Neutral Group	Retailer, wholesaler/aggregator, project developer	Afforestation and reforestation plantation, afforestation and reforestation mix native, avoided deforestation/management, energy efficiency, off-grid renewable energy	www.climateneutralgroup.com

Contd...

Name	Organization type	Project type	Website
Climate Stewards	Retailer, project developer	Afforestation and reforestation mix native	www.climatestewards.org.uk
Climat Mundi	Retailer, wholesaler/ aggregator, broker, project developer	Avoided deforestation/ management; methane: landfill	www.climatmundi.fr/Ing_EN_srub_3-Home.html
The Conservation Fund	Project developer, retailer	Afforestation and reforestation	www.conservationfund.org
Conservation International	Wholesaler/ aggregator, project developer	Afforestation and reforestation mix native, avoided deforestation/ management	www.conservation.org/xp/CIWEB
Ducks Unlimited Inc.	Project developer	Afforestation and reforestation land use	www.ducks.org
Emergent Ventures India	Broker	Afforestation and reforestation plantation; methane: livestock; energy efficiency; off-grid renewable energy	http://emergent-ventures.com
Environmental Synergy	Project developer	Afforestation and reforestation mix native	www.environmental-synergy.com/flash.html
ERA Ecosystem Restoration Associates Inc.	Retailer, project developer	Afforestation and reforestation mix native	www.econeutral.com
Greenhouse Balanced	Wholesaler/ aggregator	Afforestation and reforestation mix native	www.greenhousebalanced.com
Love Trees	Retailer, wholesaler/ aggregator, broker, project developer	Afforestation and reforestation	www.lovetrees.ca
Native Energy	Retailer, wholesaler/ aggregator	Avoided deforestation/ management; methane: livestock, landfill; renewable energy credits	www.nativeenergy.com
The Nature Conservancy	Project developer	Afforestation and reforestation plantation, avoided deforestation/ management	www.nature.org
New Forests	Wholesaler/ aggregator, project developer	Afforestation and reforestation, avoided deforestation	www.newforests.com.au
Offsetters Climate Neutral Society	Retailer, broker, project developer	Avoided deforestation/ management, energy efficiency, off-grid renewable energy	www.offsetters.com

Contd...

Name	Organization type	Project type	Website
Prima Klima	Retailer; other: fund raising and working with project developers	Afforestation and reforestation mix native	www.prima-klimaweltweit.de
SKG Sangha	Project developer	Avoided deforestation/ management; methane: livestock	www.skgsangha.org
Sterling Planet, Inc.	Retailer	Afforestation/reforestation mix native, energy efficiency, renewable energy credits	www.sterlingplanet.com
TreeBanking Inc.	Retailer	Afforestation and reforestation	www.treebankinginc.com/Home/tabid/37/ctl/Privacy/Default.aspx
Treeflights.com	Retailer	Afforestation and reforestation mix native	www.treeflights.com
The Trust for Public Land	Project developer	Afforestation and reforestation mix native	www.tpl.org
Woodland Trust		Afforestation and reforestation mix native	www.woodlandtrust.org.uk

Note: These organizations work in different roles and facilitate different aspects of the forestry sector
Source: Adapted from Hamilton, Ricardo, Guy, *et al.* (2007)

Appendix 2

PROCEDURES TO DEFINE THE ELIGIBILITY OF LANDS FOR AFFORESTATION AND REFORESTATION PROJECT ACTIVITIES*

1. Project participants shall provide evidence that the land within the planned project boundary is eligible as an afforestation and reforestation Clean Development Mechanism (CDM) project activity following the steps outlined as follows.
   a. Demonstrate that the land at the moment the project starts is not forest by providing information that:
      - the land is below the forest national thresholds (crown cover, tree height, and minimum land area) for forest definition under decisions 11/CP.7 and 19/CP.9 as communicated by the respective Designated National Authority (DNA); and
      - the land is not temporarily unstocked as a result of human intervention such as harvesting or natural causes or is not covered by young natural stands or plantations, which have yet to reach a crown density or tree height in accordance with national thresholds and which have the potential to revert to forest without human intervention.
   b. Demonstrate that the activity is a reforestation or afforestation project activity:
      - For reforestation project activities, demonstrate that on 31 December 1989, the land was below the forest national thresholds (crown cover, tree height, and minimum land area) for forest definition under decision 11/CP.7 as communicated by the respective DNA.

* CDM Executive Board Report 22, Annex 16, page 1

- For afforestation project activities, demonstrate that the land is below the forest national thresholds (crown cover, tree height, and minimum land area) for forest definition under decision 11/CP.7 as communicated by the respective DNA, for a period of at least 50 years.

2. In order to demonstrate steps 1(a) and 1(b), project participants shall provide one of the following verifiable information:
   - Aerial photographs or satellite imagery complemented by ground reference data; or
   - Ground-based surveys (land-use permits, land-use plans, or information from local registers such as cadastre, owners register, land-use or land management register); or
   - If options (a) and (b) are not available/applicable, project participants shall submit a written testimony that was produced by following a participatory rural appraisal methodology.[1]

[1] Participatory rural appraisal is an approach to the analysis of local problems and the formulation of tentative solutions with local stakeholders. It makes use of a wide range of visualization methods for group-based analysis to deal with spatial and temporal aspects of social and environmental problems.

Appendix 3

PDD Format for Large-scale Afforestation and Reforestation Projects*

CDM PDD FORM FOR AFFORESTATION AND REFORESTATION PROJECT ACTIVITIES (CDM-AR-PDD) VERSION 04

Section A: General Description of the Proposed Afforestation and Reforestation CDM Project Activity

A.1 Title of the proposed afforestation and reforestation CDM project activity

A.2 Description of the proposed afforestation and reforestation CDM project activity

A.3 Project participants

Please list project participants and party (parties) involved according to the format in Table A3.1 and provide contact information in Annex 3.1.

Table A3.1 Format of information on project participants and parties involved

Name of party involved[a] (host indicates a host party)	Private and/or public entity (entities) project participants (*) (as applicable)	Indicate if the party involved wishes to be considered a project participant (Yes/No)
Name A (host)	Private entity A	
Public entity A ...	No	
Name B	None	Yes
Name C	None	No
—	—	—

[a] In accordance with the CDM afforestation and reforestation modalities and procedures, at the time of making the CDM-AR-PDD public at the stage of validation, a party involved may or may not have provided its approval. At the time of requesting registration, the approval by the party (parties) involved is required.

Note: When the CDM-AR-PDD is prepared to support a proposed new baseline and monitoring methodology (form CDM-AR-NM), at least the host party (parties) and any known project participant (for example, those proposing a new methodology) shall be identified.

* Details available at <http://cdm://unfccc.int/Reference/PDDs_Forms/PDDs/index.html>

A.4 Description of location and boundaries of the afforestation and reforestation CDM project activity

A.4.1 Location of the proposed afforestation and reforestation CDM project activity

A.4.1.1 Host party (parties):

A.4.1.2 Region/state/province:

A.4.1.3 City/town/community:

A.4.2 Detailed geographic delineation of the project boundary, including information allowing the unique identification(s) of the proposed afforestation and reforestation CDM project activity

A.5 Technical description of the afforestation and reforestation CDM project activity

A.5.1 Description of the present environmental conditions of the area planned for the proposed afforestation and reforestation CDM project activity, including a concise description of climate, hydrology, soils, and ecosystems (including land use)

A.5.2 Description of the presence, if any, of rare or endangered species and their habitats

A.5.4 Species and varieties selected for the proposed afforestation and reforestation CDM project activity

A.5.4 Technology to be employed by the proposed afforestation and reforestation CDM project activity

A.5.5 Transfer of technology/know-how by the proposed afforestation and reforestation CDM project activity, if applicable

A.5.6 Proposed measures to be implemented to minimize potential leakage

A.6 Description of legal title to the land, current land tenure and rights to temporary certified emission reductions (tCERs)/long-term CERs (lCERs) issued for the proposed afforestation and reforestation CDM project activity

A.7 Assessment of the eligibility of the land

A.8 Approach for addressing non-permanence

A.9 Estimated amount of net anthropogenic greenhouse gas (GHG) removals by sinks over the chosen crediting period

Table A3.2 Summary of results obtained in C.7, D.1, and D.2

Year	Estimation of baseline net GHG removals by sinks (tonnes of CO_2 equivalent)	Estimation of actual net GHG removals by sinks (tonnes of CO_2 equivalent)	Estimation of leakage (tonnes of CO_2 equivalent)	Estimation of net anthropogenic GHG removals by sinks (tonnes of CO_2 equivalent)
Year A				
Year B				
Year C				
Year ...				
Total (tonnes of CO_2 equivalent)				

GHG - greenhouse gas

A.10 Public funding of the proposed afforestation and reforestation CDM project activity

SECTION B: DURATION OF THE PROJECT ACTIVITY/CREDITING PERIOD

B.1 Starting date of the proposed afforestation and reforestation CDM project activity and of the crediting period

B.2 Expected operational lifetime of the proposed afforestation and reforestation CDM project activity

B.3 Choice of crediting period

B.3.1 Length of the renewable crediting period (in years and months), if selected

B.3.2 Length of the fixed crediting period (in years and months), if selected

SECTION C: APPLICATION OF AN APPROVED BASELINE AND MONITORING METHODOLOGY

C.1 Title and reference of the approved baseline and monitoring methodology applied to the proposed afforestation and reforestation CDM project activity

C.2 Assessment of the applicability of the selected approved methodology to the proposed afforestation and reforestation CDM project activity and justification of the choice of the methodology

C.3 Assessment of the selected carbon pools and emission sources of the approved methodology to the proposed CDM project activity

C.4 Description of strata identified using the ex-ante stratification

C.5 Identification of the baseline scenario

C.5.1 Description of the application of the procedure to identify the most plausible baseline scenario (separately for each stratum defined in C.4)

C.5.2 Description of the identified baseline scenario (separately for each stratum defined in C.4)

C.6. Assessment and demonstration of additionality

C.7 Estimation of the ex-ante baseline net GHG removals by sinks

Table A3.3 Details of data collected for project activity

ID number[a]	Data variable	Data unit	Value applied	Data source	Comment

[a] Please provide ID number for cross-referencing in the project design document

Table A3.4 Tabular format for presenting the final results of calculations

Year	Annual estimation of baseline net anthropogenic GHG removals by sinks in tonnes of CO_2 equivalent
Year A	
Year B	
Year C	
Year ...	
Total estimated baseline net GHG removals by sinks (tonnes of CO_2 equivalent)	
Total number of crediting years	
Annual average over the crediting period of estimated baseline net GHG removals by sinks (tonnes of CO_2 equivalent)	

GHG – greenhouse gas

C.8 Date of completion of the baseline study and the name of person (persons)/entity (entities) determining the baseline

SECTION D: ESTIMATION OF EX-ANTE ACTUAL NET GHG REMOVALS BY SINKS, LEAKAGE, AND ESTIMATED AMOUNT OF NET ANTHROPOGENIC GHG REMOVALS BY SINKS OVER THE CHOSEN CREDITING PERIOD

D.1 Estimate of the ex-ante actual net GHG removals by sinks

D.2 Estimate of the ex-ante leakage

SECTION E: MONITORING PLAN

E.1 Monitoring of the project implementation

E.1.1 Monitoring of forest establishment and management

Table A3.5 Details of data collected for project activity

ID number[a]	Data variable	Data unit	Measured (m), calculated (c), estimated (e), or default (d)[b]	Recording frequency	Number of data points/other measure of number of collected data	Comment

[a] Please provide the ID number for cross-referencing in the project design document.

[b] Please provide full reference to data source.

E.1.2 If required by the selected approved methodology, describe or provide reference to standard operating procedures (SOPs) and quality control/quality assurance (QA/QC) procedures applied.

Table A3.6 Details of data collected for project activity

ID number[a]	Data variable	Data unit	Measured (m), calculated (c), estimated (e), or default (d)[b]	Recording frequency	Number of data points/other measure of number of collected data	Comment

[a] Please provide the ID number for cross-referencing in the project design document.

[b] Please provide full reference to data source.

E.2 Sampling design and stratification

E.3 Monitoring of the baseline net GHG removals by sinks, if required by the selected approved methodology

Table A3.7 Details of data collected for project activity

ID number[a]	Data variable	Data unit	Measured (m), calculated (c), estimated (e), or default (d)[b]	Recording frequency	Number of sample plots at which the data will be monitored	Comment

[a] Please provide the ID number for cross-referencing in the project design document.

[b] Please provide full reference to data source.

E.4 Monitoring of the actual net GHG removals by sinks

E.4.1 Data to be collected to monitor the verifiable changes in carbon stock in the carbon pools within the project boundary resulting from the proposed afforestation and reforestation CDM project activity

Table A3.8 Details of data collected for project activity

ID number[a]	Data variable	Data unit	Measured (m), calculated (c), estimated (e), or default (d)[b]	Recording frequency	Number of sample plots at which the data will be monitored	Comment

[a] Please provide the ID number for cross-referencing in the project design document.

[b] Please provide full reference to data source.

E.4.2 Data to be collected to monitor the GHG emissions by the sources, measured in units of CO_2 equivalent, that are increased as a result of the implementation of the proposed afforestation and reforestation CDM project activity within the project boundary

Table A3.9 Details of data collected for project activity

ID number[a]	Data variable	Data unit	Measured (m), calculated (c), estimated (e), or default (d)[b]	Recording frequency	Number of sample plots at which the data will be monitored	Comment

[a] Please provide the ID number for cross-referencing in the project design document.
[b] Please provide full reference to data source.

E.5 Leakage

E.5.1 If applicable, please describe the data and information that will be collected to monitor the leakage of the proposed afforestation and reforestation CDM project activity.

Table A3.10 Details of data collected for project activity

ID number[a]	Data variable	Data unit	Measured (m), calculated (c), estimated (e), or default (d)[b]	Recording frequency	Number of data points	Comment

[a] Please provide the ID number for cross-referencing in the project design document.
[b] Please provide full reference to data source.

E.5.2 Specify the procedures for the periodic review of implementation of activities and measures to minimize leakage, if required, by the selected approved methodology.

E.6 Provide any additional QC and QA procedures undertaken for data monitored not included in E.1.3.

Table A3.11 Details of data collected for project activity

Data (indicate ID number)	Uncertainty level of data (high/medium/low)	Explain QA/QC procedures planned for these data, or why such procedures are not necessary

QA – quality assurance; QC – quality control

E.7 Please describe the operational and management structure(s) that the project operator will implement to monitor actual GHG removals by sinks and any leakage generated by the proposed afforestation and reforestation CDM project activity.

E.8 Name of person (persons)/entity (entities) applying the monitoring plan.

SECTION F: ENVIRONMENTAL IMPACTS OF THE PROPOSED AFFORESTATION AND REFORESTATION CDM PROJECT ACTIVITY

F.1 Documentation on the analysis of the environmental impacts, including impacts on biodiversity and natural ecosystems, and impacts outside the project boundary of the proposed afforestation and reforestation CDM project activity

F.2 If any negative impact is considered significant by the project participants or the host party, a statement that project participants have undertaken an environmental impact assessment, in accordance with the procedures required by the host party, including conclusions and all references to support documentation

F.3 Description of planned monitoring and remedial measures to address significant impacts referred to in F.2

SECTION G: SOCIO-ECONOMIC IMPACTS OF THE PROPOSED AFFORESTATION AND REFORESTATION CDM PROJECT ACTIVITY

G.1 Documentation on the analysis of the major socio-economic impacts, including impacts outside the project boundary of the proposed afforestation and reforestation CDM project activity

G.2 If any negative impact is considered significant by the project participants or the host party, a statement that project participants have undertaken a socio-economic impact assessment, in accordance with the procedures required by the host party, including conclusions and all references to supporting documentation

G.3 Description of planned monitoring and remedial measures to address significant impacts referred to in G.2

SECTION H: STAKEHOLDERS' COMMENTS

H.1 Brief description of how comments by local stakeholders have been invited and compiled

H.2 Summary of the comments received

H.3 Report on how due account was taken of any comments received

Annex 3.1: Contact Information on Participants in the Proposed Afforestation and Reforestation CDM Project Activity

Table A3.12 Contact details of participants in the CDM project activity

Organization	
Street/PO Box	
Building	
City	
State/region	
Postfix/ZIP	
Country	
Telephone	
FAX	
E-Mail	
URL	
Represented by	
Title	
Salutation	
Last Name	
Middle Name	
First Name	
Department	
Mobile	
Direct FAX	
Direct telephone	
Personal e-mail	

Annex 3.2: Information Regarding Public Funding

Annex 3.3: Baseline Information

Annex 3.4: Monitoring Plan

Table A3.13 History of revisions in the project design document

Version	Date	Nature of revision
04	EB35, Annex 20 19 October 2007	• Restructuring of Section A • Section "Monitoring of forest establishment and management" replaces sections "Monitoring of the project boundary", and "Monitoring of forest management" • Introduced a new section allowing for explicit description of SOPs and QA/QC procedures if required by the selected approved methodology • Change in design of the section "Monitoring of the baseline net GHG removals by sinks" allowing for more efficient presentation of data
03	EB26, Annex 19 29 September 2006	Revisions in different sections to reflect equivalent forms used by the Meth Panel and assist in making more transparent the selection of an approved methodology for a proposed afforestation and reforestation CDM project activity.
02	EB23, Annex 15a/b 24 February 2006	Inclusion of a section on the assessment of the eligibility of land and the sampling design and stratification during monitoring
01	EB15, Annex 6 03 September 2004	Initial adoption

CDM – Clean Development Mechanism; GHG – greenhouse gas; QA – quality assurance; QC – quality control

Appendix 4

List of Registered CDM Forestry Projects Till 2010

Name of project	Country	Date of registration
Assisted Natural Regeneration of Degraded Lands in Albania	Albania	02 January 2010
Carbon Sequestration through Reforestation in the Bolivian Tropics by Smallholders of "The Federación de Comunidades Agropecuarias de Rurrenabaque (FECAR)"	Bolivia	11 June 2009
Reforestation as Renewable Source of Wood Supplies for Industrial Use in Brazil	Brazil	21 July 2010
Nerquihue Small-scale CDM Afforestation Project using Mycorrhizal Inoculation in Chile	Chile	27 May 2010
Facilitating Reforestation for Guangxi Watershed Management in Pearl River Basin	China	10 November 2006
Afforestation and Reforestation on Degraded Lands in Northwest Sichuan, China	China	16 November 2009
Reforestation on Degraded Lands in Northwest Guangxi	China	15 September 2010
Forestry Project for the Basin of the Chinchiná River, an Environmental and Productive Alternative for the City and the Region	Colombia	16 April 10
Humbo Ethiopia Assisted Natural Regeneration Project	Ethiopia	07 December 2009
Reforestation of Severely Degraded Landmass in Khammam District of Andhra Pradesh, India under ITC Social Forestry Project	India	05 June 2009
Small Scale Cooperative Afforestation CDM Pilot Project Activity on Private Lands Affected by Shifting Sand Dunes in Sirsa, Haryana	India	23 March 2009
The International Small Group and Tree Planting Programme (TIST), Tamil Nadu, India	India	15 January 2010

Contd...

Name of project	Country	Date of registration
Moldova Soil Conservation Project	Moldova	30 January 2009
Reforestation of Croplands and Grasslands in Low Income Communities of Paraguarí Department, Paraguay	Paraguay	06 September 2009
Reforestation, sustainable production and carbon sequestration project in José Ignacio Távara´s dry forest, Piura, Peru	Peru	16 November 2009
Uganda Nile Basin Reforestation Project No. 3	Uganda	21 August 2009
Cao Phong Reforestation Project	Vietnam	28 April 2009

CDM – Clean Development Mechanism

Appendix 5

LIST OF MAJOR CARBON CREDITS CERTIFICATION STANDARDS/PROGRAMMES

Standard/programme	Description
CCB Standard	The Climate, Community, and Biodiversity (CCB) Standard was created to support land management projects that sequester greenhouse gas (GHG), support sustainable development, and conserve biodiversity.
Chicago Climate Exchange	The Chicago Climate Exchange (CCX) sets standards for the listing of credits (including those from offset projects) on their exchange, including the requirement of a CCX-approved third-party verifier.
Climate Neutral Network	The Climate Neutral Network functions as a hybrid between a consultant, a certifier, and a retailer and helps in branding products and companies with its Climate Cool logo.
Green-e Energy	Green-e Energy is the most widely accepted certifier of renewable energy certificates (RECs) in the USA and has a third-party verification standard for RECs as carbon offsets.
ISO 14064 Standard	The goal of this standard is "to provide a set of unambiguous and verifiable requirements" for projects of all kinds.
Voluntary Carbon Standard	The Voluntary Carbon Standard aims at providing a basic quality threshold for all carbon credits in the voluntary market.
The Gold Standard	The Gold Standard seeks to define quality for the high-end market of carbon credits.
WRI/WBCSD Protocol	Unlike the other standards described above, the World Resources Institute (WRI)/World Business Council for Sustainable Development (WBCSD) GHG Protocol is an accounting guide that does not certify offsets or organizations. However, the protocol now plays a major role in setting guidelines for GHG reporting, and helps companies in designing and implementing GHG programmes.
Plan Vivo	Plan Vivo is a system for developing community-based payments for ecosystem services projects and programmes, including agroforestry projects.

Contd...

Standard/programme	Description
Environmental Protection Agency	The Environmental Protection Agency established the Climate Action Reserve to facilitate the formulation of policies and protocols for carbon offset projects and third-party certification. It has a forest project protocol that includes reforestation (10 years), improved forest management, and avoided conversion (100-year commitment).
Climate Action Reserve	Climate Action Reserve is the parent organization of the California Registry, which tracks and registers voluntary projects that reduce emissions of GHGs.
National Carbon Offset Coalition	Founded in 2001, the sole focus of the National Carbon Offset Coalition is to help farmers, ranchers, private forest owners, and tribal and state governments tap into a new revenue stream—the sale of verified carbon credits derived from their land through trading on the CCX.
California Climate Action Registry	The California Climate Action Registry is a programme of the Climate Action Reserve and serves as a voluntary GHG registry to protect and promote early actions to reduce GHG emissions by organizations. The California Registry has provided leadership on climate change by developing and promoting credible, accurate, and consistent GHG reporting standards and tools for organizations to measure and monitor third-party verification and to reduce their GHG emissions consistently across industry sectors and geographical borders.
Climate Registry	The Climate Registry is a non-profit collaboration among North American states, provinces, territories, and Native Sovereign Nations that sets consistent and transparent standards to calculate, verify, and publicly report GHG emissions into a single registry.
National Voluntary Greenhouse Gas Reporting Programme	This system is now closed.
Scientific Certification Systems	Founded in 1984, the Scientific Certification Systems is a leading third-party provider of certification, auditing and testing services, and standards. Its goal is to recognize the highest levels of performance in food safety and quality, environmental protection and social responsibility in the private and public sectors, and to stimulate continuous improvement in sustainable development.
Renewable Energy Certificate System (Europe)	The Renewable Energy Certificate System was created to stimulate international renewable energy development. The system advocates a standard certificate as evidence of the production of a standard renewable energy quantity and provides a methodology that enables renewable energy trade. This enables a market for renewable energy to be created and thus promoting the development of new renewable energy capacity in Europe.

Appendix 6

A List of Regulatory Documents

This appendix contains the list of regulatory documents applicable to afforestation and reforestation Clean Development Mechanism (CDM) project activities. Regulatory documents can be classified into the following hierarchical categories. The documents have been assigned numbers for facilitating the user (UNFCCC 2013).

C.1 CMP DECISIONS

Readers interested in high-level source documents on CDM may consult the following documents:

C.1.1 Kyoto Protocol

C.1.2 Decision 3/CMP.1: Modalities and procedures for CDM as defined in Article 12 of the Kyoto Protocol

C.1.3 Decision 5/CMP.1: Modalities and procedures for afforestation and reforestation project activities under CDM in the first commitment period of the Kyoto Protocol

C.1.4 Decision 6/CMP.1: Simplified modalities and procedures for small-scale afforestation and reforestation project activities under the CDM in the first commitment period of the Kyoto Protocol and measures to facilitate their implementation

Readers interested in the development of afforestation and reforestation CDM project activities should consult the following documents:

C.2 CDM STANDARDS

C.2.1 Clean Development Mechanism project standard

C.2.2 Clean Development Mechanism validation and verification standard

C.2.3 Clean Development Mechanism project cycle procedure

C.3 AFFORESTATION AND REFORESTATION CDM METHODOLOGIES

C.3.1 Afforestation and reforestation of degraded mangrove habitats

C.3.2 Afforestation and reforestation of lands except wetlands

C.4 SMALL-SCALE AFFORESTATION AND REFORESTATION CDM METHODOLOGIES

C.4.1 Simplified baseline and monitoring methodology for small-scale CDM afforestation and reforestation project activities implemented on wetlands

C.4.2 Simplified baseline and monitoring methodology for small-scale CDM afforestation and reforestation project activities implemented on lands other than wetlands

C.5 AFFORESTATION AND REFORESTATION CDM METHODOLOGICAL TOOLS

C.5.1 Combined tool to identify the baseline scenario and demonstrate additionality in afforestation and reforestation CDM project activities

C.5.2 Demonstrating eligibility of land for afforestation and reforestation CDM project activities

C.5.3 Estimation of carbon stocks and change in carbon stocks of trees and shrubs in afforestation and reforestation CDM project activities

C.5.4 Estimation of carbon stocks and change in carbon stocks in dead wood and litter in afforestation and reforestation CDM project activities

C.5.5 Tool for the estimation of change in soil organic carbon stocks due to the implementation of afforestation and reforestation CDM project activities

C.5.6 Demonstrating appropriateness of allometric equations for estimation of above-ground tree biomass in afforestation and reforestation CDM project activities

C.5.7 Demonstrating appropriateness of volume equations for the estimation of above-ground tree biomass in afforestation and reforestation CDM project activities

C.5.8 Estimation of non-CO_2 GHG emissions resulting from burning of biomass attributable to an afforestation and reforestation CDM project activity

C.5.9 Estimation of the increase in GHG emissions attributable to the displacement of pre-project agricultural activities in afforestation and reforestation CDM project activity

C.5.10 Calculation of the number of sample plots for measurements within afforestation and reforestation CDM project activities

C.6 AFFORESTATION AND REFORESTATION CDM METHODOLOGICAL GUIDELINES

C.6.1 Establishment of standardized baselines for afforestation and reforestation project activities under the CDM

C.6.2 Guidelines for completing the proposed new afforestation and reforestation baseline and monitoring methodology form

C.6.3 Guidelines on accounting of specified types of changes in afforestation and reforestation CDM project activities from the description in registered project design documents

C.7 AFFORESTATION AND REFORESTATION CDM FORMS

C.7.1 F-CDM-AR-PDD: Project design document form for afforestation and reforestation CDM project activities

C.7.2 F-CDM-SSC-AR-PDD: Project design document form for small-scale afforestation and reforestation CDM project activities

C.7.3 F-CDM-AR-PoA-DD: Programme design document form for afforestation and reforestation CDM programmes of activities

C.7.4 F-CDM-SSC-AR-PoA-DD: Programme design document form for small-scale afforestation and reforestation CDM programmes of activities

C.7.5 F-CDM-AR-CPA-DD: Component project activity design document form for afforestation and reforestation component project activities

C.7.6 F-CDM-SSC-AR-CPA-DD: Component project activity design document form for small-scale afforestation and reforestation component project activities

C.7.7 CDM-AR-NM-FORM: Proposed new afforestation and reforestation baseline and monitoring methodology form

Appendix 7

METHODOLOGIES FOR AFFORESTATION AND REFORESTATION CDM PROJECT ACTIVITIES

INTRODUCTION

Project participants willing to validate/register an afforestation and reforestation CDM project activity shall:

- use an afforestation and reforestation methodology previously approved by the Executive Board, or
- propose a new afforestation and reforestation methodology to the Executive Board for consideration and approval, if appropriate.

Approved Afforestation and Reforestation Methodologies

This page provides you with a list of baseline and monitoring methodologies for afforestation and reforestation project activities approved by the CDM Executive Board.

Proposed New Afforestation and Reforestation Methodologies under Consideration

This page provides you with baseline and monitoring methodologies for afforestation and reforestation project activities currently under consideration by the CDM Executive Board.

How to Propose a New Afforestation and Reforestation Methodology

This page provides you with information about the procedures for submission of proposed new afforestation and reforestation methodologies and the next deadline for submission.

What is a Methodology?

There are two types of methodologies for afforestation and reforestation (afforestation and reforestation) projects: baseline methodology and monitoring methodology.

Baseline methodology: In the context of afforestation and reforestation projects, baseline methodology can be defined as follows:

"A methodology is an application of an approach, as defined in paragraph 22 of the CDM afforestation and reforestation modalities and procedures, to an individual afforestation and reforestation CDM project activity, for the determination of the baseline scenario."

A baseline methodology should reflect aspects such as environmental conditions and past land uses and land-use changes. No methodology is excluded a priori so that project participants have the opportunity to propose a methodology. In considering paragraph 22, the Executive Board agreed that the following cases apply:

- **Case of a new methodology:** In developing a baseline methodology, the first step is to identify the most appropriate approach for the proposed afforestation and reforestation CDM project activity and then an applicable methodology.
- **Case of an approved methodology:** In opting for an approved methodology, project participants have implicitly chosen an approach (Glossary of CDM Terms, Version 03).

Ultimately, the baseline methodology used in a CDM project (including an afforestation and reforestation project) represents the means by which the baseline scenario for that project is defined. The baseline scenario for afforestation and reforestation projects is in turn defined as the scenario that reasonably represents the sum of the changes in carbon stocks in the carbon pools within the project boundary that would occur in the absence of the proposed afforestation and reforestation CDM project activity [EB 35 Annex 21, Part III, paragraph 10 (a)].

Thus, afforestation and reforestation baseline methodologies are the tools used to define the changes in carbon stocks in the carbon pools within the project boundary that would occur in the absence of the proposed afforestation and reforestation project.

Monitoring methodology: In the context of afforestation and reforestation projects, monitoring methodology can be defined as follows:

"A monitoring methodology refers to the method used by project participants for the collection and archiving of all relevant data necessary for the implementation of the monitoring plan (Glossary of CDM Terms, Version 03)."

Therefore, a monitoring methodology is the means by which the net anthropogenic GHG removals by sinks achieved by the afforestation and reforestation project are calculated, taking into account any emissions from sources within the project boundary. A monitoring methodology sets out how project proponents should develop and implement a monitoring plan for a project to gather the data required for calculating emission removals generated.

Both the baseline methodology and the monitoring methodology must be specified in the project design document (PDD). Project participants may use the existing baseline and monitoring methodologies if methodologies appropriate to the project already exist. In this event, the PDD need only state the baseline and monitoring methodologies to be used (by reference number). If no baseline or monitoring methodologies appropriate to the project have yet been developed, a new methodology must be developed. This section explains the process for the development, submission, approval, and revision of methodologies for use in afforestation and reforestation projects.

Baseline and monitoring methodologies for afforestation and reforestation projects must be developed together, using a single form (CDM-AR-NM). The Executive Board has clarified that once a methodology is approved, it may be used for both CDM project activities and CDM programme activities under a programme of activities (EB 35, paragraph 15).

A list of approved afforestation and reforestation methodologies and proposed afforestation and reforestation methodologies under consideration is available.

Source: UNFCCC website

APPROVED AFFORESTATION AND REFORESTATION METHODOLOGIES

This section provides access to approved methodologies and the methodological tools agreed by the Executive Board. Some of the approved afforestation and reforestation tools are as follows:

- Tool for the demonstration and assessment of additionality in afforestation and reforestation CDM project activities.
- Combined tool to identify the baseline scenario and to demonstrate additionality in afforestation and reforestation CDM project activities.
- Calculation of the number of sample plots for measurements within afforestation and reforestation CDM project activities.
- Tool for testing significance of GHG emissions in afforestation and reforestation CDM project activities.
- Estimation of GHG emissions related to fossil fuel combustion in afforestation and reforestation CDM project activities.
- Procedure to determine when accounting of the soil organic carbon pool may be conservatively neglected in CDM afforestation and reforestation project activities.
- Estimation of direct nitrous oxide emission from nitrogen fertilization.
- Tool for estimation of GHG emissions from clearing, burning, and decay of existing vegetation due to implementation of a CDM afforestation and reforestation project activity.
- Tool for estimation of GHG emissions related to displacement of grazing activities in afforestation and reforestation CDM project activity.[1]
- Tool for calculating GHG emissions due to leakage from increased use of non-renewable woody biomass attributable to an afforestation and reforestation CDM project activity.
- Tool for estimation of carbon stocks, and removal and emission for the dead organic matter pool due to the implementation of a CDM afforestation and reforestation project activity.
- Tool for the identification of degraded or degrading lands for consideration in implementing CDM afforestation and reforestation project activities.
- Estimation of changes in the carbon stocks of existing trees and shrubs within the boundary of an afforestation and reforestation CDM project activity.

[1] Effective 4 June 2011, this tool shall be superseded by the afforestation and reforestation tool "Estimation of the increase in GHG emissions attributable to displacement of pre-project agricultural activities in afforestation and reforestation CDM project activity" (see EB 51, para 40).

- Estimation of the increase in GHG emissions attributable to displacement of pre-project agricultural activities in afforestation and reforestation CDM project activity.
- Tool for estimating change in soil organic carbon stocks due to the implementation of afforestation and reforestation CDM project activities.

Appendix 7.A

APPROVED METHODOLOGIES FOR LARGE-SCALE AFFORESTATION AND REFORESTATION CDM PROJECTS

Table A7.1 Some approved large-scale methodologies

Methodology number	Methodology title (including baseline and monitoring methodologies)	Sectoral scope	Approval history
AR-AM0002	Restoration of degraded lands through afforestation/ reforestation (Version 3) Tools referenced in this methodology are as follows: • Tool for the demonstration and assessment of additionality in afforestation and reforestation CDM project activities. • Combined tool to identify the baseline scenario and demonstrate additionality in afforestation and reforestation CDM project activities. • Tool for the identification of degraded or degrading lands for consideration in implementing CDM afforestation and reforestation project activities.	14	ARNM0007-rev
AR-AM0004	Reforestation or afforestation of land currently under agricultural use (Version 4) Tool referenced in this methodology is as follows: • Tool for the demonstration and assessment of a dditionality in afforestation and reforestation CDM project activities.	14	ARNM0019
AR-AM0005	Afforestation and reforestation project activities implemented for industrial and/or commercial uses (Version 4) Tools referenced in this methodology are as follows: • Tool for the demonstration and assessment of additionality in afforestation and reforestation CDM project activities.	14	ARNM0015-rev

Contd...

Table A7.1 *Contd...*

Methodology number	Methodology title (including baseline and monitoring methodologies)	Sectoral scope	Approval history
	• Combined tool to identify the baseline scenario and demonstrate additionality in afforestation and reforestation CDM project activities. • Calculation of the number of sample plots for measurements within afforestation and reforestation CDM project activities. • Tool for testing the significance of GHG emissions in afforestation and reforestation CDM project activities. • Tool for the estimation of GHG emissions related to displacement of grazing activities in afforestation and reforestation CDM project activities. • Tool for the identification of degraded or degrading lands for consideration in implementing CDM afforestation and reforestation project activities.		
AR-AM0006	Afforestation and reforestation with trees supported by shrubs on degraded land (Version 3) referenced in this methodology are as follows: • Tool for the demonstration and assessment of additionality in afforestation and reforestation CDM project activities. • Combined tool to identify the baseline scenario and demonstrate additionality in afforestation and reforestation CDM project activities.	14	ARNM0020-rev
AR-AM0007	Afforestation and reforestation of land currently under agricultural or pastoral use (Version 5) Tools referenced in this methodology are as follows: • Combined tool to identify the baseline scenario and demonstrate additionality in afforestation and reforestation CDM project activities. • Procedure to determine when the accounting of the soil organic carbon pool may be conservatively neglected in CDM afforestation and reforestation project activities. • Calculation of the number of sample plots for measurements within afforestation and reforestation CDM project activities. • Tool for testing the significance of GHG emissions in afforestation and reforestation CDM project activities.	14	ARNM0021-rev

Contd...

Table A7.1 *Contd...*

Methodology number	Methodology title (including baseline and monitoring methodologies)	Sectoral scope	Approval history
AR-AM0009	Afforestation or reforestation on degraded land allowing for silvopastoral activities (Version 4) Tools referenced in this methodology are as follows: • Combined tool to identify the baseline scenario and demonstrate additionality in afforestation and reforestation CDM project activities. • Calculation of the number of sample plots for measurements within afforestation and reforestation CDM project activities. • Tool for testing the significance of GHG emissions in afforestation and reforestation CDM project activities. • Tool for the identification of degraded or degrading lands for consideration in implementing CDM afforestation and reforestation project activities. • Tool for testing the significance of GHG emissions in afforestation and reforestation CDM project activities.	14	ARNM0024-rev
AR-AM0010	Afforestation and reforestation project activities implemented on unmanaged grasslands in reserved/ protected areas (Version 4) Tools referenced in this methodology are as follows: • Tool for the demonstration and assessment of additionality in afforestation and reforestation CDM project activities. • Calculation of the number of sample plots for measurements within afforestation and reforestation CDM project activities. • Procedure to determine when accounting of the soil organic carbon pool may be conservatively neglected in CDM afforestation and reforestation project activities.	14	ARNM0034
AR-AM0011	Afforestation and reforestation of land subject to polyculture farming (Version 1) (520 KB) Tools referenced in this methodology are as follows: • Combined tool to identify the baseline scenario and demonstrate additionality in afforestation and reforestation CDM project activities. • Calculation of the number of sample plots for measurements within afforestation and reforestation CDM project activities.	14	ARNM0036

Contd...

Table A7.1 *Contd...*

Methodology number	Methodology title (including baseline and monitoring methodologies)	Sectoral scope	Approval history
	• Tool for the estimation of GHG emissions from clearing, burning, and decay of existing vegetation due to implementation of CDM afforestation and reforestation project activities. • Tool for the identification of degraded or degrading lands for consideration in implementing CDM afforestation and reforestation project activities. • Estimation of the increase in GHG emissions attributable to displacement of pre-project agricultural activities in afforestation and reforestation CDM project activities.		

CDM – Clean Development Mechanism; GHG – greenhouse gas

Note: Methodologies approved till May 2010

Source: Details available at <http://cdm.unfccc.int/methodologies/ARmethodologies/approved_ar.html>

Table A7.2 Some approved consolidated methodologies

Methodology number	Methodology title (including baseline and monitoring methodologies)	Sectoral scope	Consolidated sources
AR-ACM0001	Afforestation and reforestation of degraded land (Version 4) Tools referenced in this methodology are as follows: • Combined tool to identify the baseline scenario and demonstrate additionality in afforestation and reforestation CDM project activities. • Calculation of the number of sample plots for measurements within afforestation and reforestation CDM project activities. • Tool for testing the significance of GHG emissions in afforestation and reforestation CDM project activities. • Tool for the estimation of GHG emissions from clearing, burning, and decay of existing vegetation due to implementation of CDM afforestation and reforestation project activities. • Tool for the estimation of GHG emissions related to displacement of grazing activities in afforestation and reforestation CDM project activities. • Tool for the identification of degraded or degrading lands for consideration in implementing CDM afforestation and reforestation project activities.	14	ARNM0032-rev Replaces: AR-AM0003
AR-ACM0002	Afforestation or reforestation of degraded land without displacement of pre-project activities—Version 1 Tools referenced in this methodology are as follows: • Calculation of the number of sample plots for measurements within afforestation and reforestation CDM project activities.	14	Replaces: AR-AM0001 AR-AM0008

Contd...

Table A7.2 *Contd...*

Methodology number	Methodology title (including baseline and monitoring methodologies)	Sectoral scope	Consolidated sources
	• Tool for testing the significance of GHG emissions in afforestation and reforestation CDM project activities. • Procedure to determine when accounting of the soil organic carbon pool may be conservatively neglected in CDM afforestation and reforestation project activities. • Combined tool to identify the baseline scenario and demonstrate additionality in afforestation and reforestation CDM project activities. • Procedures to demonstrate the eligibility of lands for afforestation and reforestation CDM project activities. • Tool for the identification of degraded or degrading lands for consideration in implementing CDM afforestation and reforestation project activities. • Tool for the estimation of GHG emissions from clearing, burning, and decay of existing vegetation due to the implementation of CDM afforestation and reforestation project activities.		

CDM – Clean Development Mechanism; GHG – greenhouse gas

Appendix 7.B

APPROVED METHODOLOGIES FOR SMALL-SCALE AFFORESTATION AND REFORESTATION CDM PROJECTS

Table A7.3 Some approved small-scale methodologies

Methodology number	Methodology title (including baseline and monitoring methodologies)	Sectoral scope	Type
AR-AMS0001	Simplified baseline and monitoring methodologies for small-scale afforestation and reforestation project activities under the CDM implemented on grasslands or croplands (Version 5)	14	
AR-AMS0002	Simplified baseline and monitoring methodologies for small-scale afforestation and reforestation project activities under the CDM implemented on settlements (Version 2)	14	
AR-AMS0003	Simplified baseline and monitoring methodology for small-scale CDM afforestation and reforestation project activities implemented on wetlands (Version 1)	14	
AR-AMS0004	Simplified baseline and monitoring methodology for small-scale agroforestry afforestation and reforestation project activities under the Clean Development Mechanism (Version 2)	14	
AR-AMS0005	Simplified baseline and monitoring methodology for small-scale afforestation and reforestation project activities under the CDM implemented on lands having low inherent potential to support living biomass (Version 2)	14	
AR-AMS0006	Simplified baseline and monitoring methodology for small-scale silvopastoral afforestation and reforestation project activities under the CDM (Version 1)	14	

CDM – Clean Development Mechanism

Source: UNFCCC

Appendix 8

FORMAT FOR SMALL-SCALE AFFORESTATION AND REFORESTATION PROJECTS

CLEAN DEVELOPMENT MECHANISM PROJECT DESIGN DOCUMENT FORM FOR SMALL-SCALE AFFORESTATION AND REFORESTATION PROJECT ACTIVITIES (CDM-SSC-AR-PDD) (VERSION 02)

Section A: General Description of the Proposed Small-scale Afforestation and Reforestation CDM Project Activity

A.1 Title of the proposed small-scale afforestation and reforestation CDM project activity

A.2 Description of the proposed small-scale afforestation and reforestation CDM project activity

A.3 Project participants

Please list the project participants and party/parties involved and provide contact information in Annex 8.1.

Table A8.1 Tabular format for listing details of project participants and parties involved

Name of party involved[a] (host indicates a host party)	Private and/or public entity/ entities project participants (as applicable)(see a)	Indicate if the party involved wishes to be considered a project participant (Yes/No)
Name A (host)	Private entity A	
Public entity A ...	No	
Name B	None	Yes
Name C	None	No
—	—	—

[a] At the time of making the CDM-SSC-AR-PDD public at the stage of validation, a party involved may or may not have provided its approval. At the time of requesting registration, the approval by the party/parties involved is required.

A.4 Description of location and boundary of the small-scale afforestation and reforestation CDM project activity

A.4.1 Location of the proposed small-scale afforestation and reforestation CDM project activity

A.4.1.1 Host party/parties

A.4.1.2 Region/state/province

A.4.1.3 City/town/community

A.4.2 Detail of geographical location and project boundary, including information allowing the unique identification(s) of the proposed small-scale afforestation and reforestation CDM project activity:

A.5 Technical description of the small-scale afforestation and reforestation CDM project activity

A.5.1 Type(s) of small-scale afforestation and reforestation CDM project activity

A.5.2 A concise description of the present environmental conditions of the area, which include information on climate, soils, main watershed, ecosystems, and the possible presence of rare or endangered species and their habitats

A.5.3 Species and varieties selected

A.5.4 Technology to be employed by the proposed small-scale afforestation and reforestation CDM project activity

A.5.5 Transfer of technology/know-how, if applicable

A.5.6 Proposed measures to be implemented to minimize potential leakage as applicable

A.6 A description of legal title to the land, current land tenure and land use, and rights to tCERs/lCERs issued

A.7 Assessment of the eligibility of land

A.8 Approach for addressing non-permanence

Please select between:

- Issuance of tCERs
- Issuance of lCERs

A.9 Duration of the proposed small-scale afforestation and reforestation CDM project activity/crediting period

A.9.1 Starting date of the proposed small-scale afforestation and reforestation CDM project activity and of the (first) crediting period, including a justification

A.9.2 Expected operational lifetime of the proposed small-scale afforestation and reforestation CDM project activity

A.9.3 Choice of crediting period and related information

Please select one of the following:

- Renewable crediting period
- Fixed Crediting period

A.9.3.1 Duration of the first crediting period (in years and months), if a renewable crediting period is selected

A.9.3.2 Duration of the fixed crediting period (in years and months), if selected

A.10 Estimated amount of net anthropogenic GHG removals by sinks over the chosen crediting period

Please provide the total estimation of net anthropogenic GHG removals by sinks as well as annual estimates for the chosen crediting period. Information on the net anthropogenic GHG removals by sinks shall be indicated using the following tabular format:

Table A8.2 Total estimation of net GHG removals by sinks

Years	Annual estimation of net anthropogenic GHG removals by sinks in tonnes of CO_2 equivalent
Year A	
Year B	
Year C	
Year ...	
Total estimated net anthropogenic GHG removals by sinks (tonnes of CO_2 equivalent)	
Total number of crediting years	
Annual average over the crediting period of estimated net anthropogenic GHG removals by sinks (tonnes of CO_2 equivalent)	

GHG – greenhouse gas

A.11 Public funding of the proposed small-scale afforestation and reforestation CDM project activity

A.12 Confirmation that the small-scale afforestation and reforestation CDM project activity is not a de-bundled component of a larger project activity

Section B: Application of a Baseline and Monitoring Methodology

B.1 Title and reference of the approved baseline and monitoring methodology applied to the proposed small-scale afforestation and reforestation CDM project activity

B.2 Justification of the applicability of the baseline and monitoring methodology to the proposed small-scale afforestation and reforestation CDM project activity

B.3 Specification of the GHGs whose emissions will be part of the proposed small-scale afforestation and reforestation CDM project activity

B.4 Carbon pools selected In calculating the baseline net GHG removals by sinks and/or actual net GHG removals by sinks, project participants may choose not to account for one or more carbon pools, and/or emissions of GHGs measured in units of CO_2 equivalents, while avoiding double counting.

Select the carbon pools that are considered in determining the actual net GHG removals by sinks and baseline net GHG removals by sinks in Table A8.3 in accordance with the proposed new/approved methodology used. Note that the same carbon pools should be considered in the actual net GHG removals by sinks and the baseline net GHG removals by sinks.

Table A8.3 Status of carbon pool selected (yes or no)

Carbon pools	Selected (answer with yes or no)
Above ground	
Below ground	
Dead wood	
Litter	
Soil organic carbon	

B.5 Description of strata applied for ex-ante estimations

B.6 Application of baseline methodology to the proposed small-scale afforestation and reforestation CDM project activity

B.7 Description of how the actual net GHG removals by sinks are increased above those that would have occurred in the absence of the registered small-scale afforestation and reforestation CDM project activity

B.8 Application of monitoring methodology and monitoring plan to the small-scale afforestation and reforestation CDM project activity

B.8.1 Data to be monitored: Monitoring of the actual net GHG removals by sinks and leakage

B.8.1.1 Actual net GHG removals by sinks data

B.8.1.1.1 Data to be collected or used to monitor the verifiable changes in carbon stock in the carbon pools within the project boundary resulting from the proposed small-scale afforestation and reforestation CDM project activity, and how this data will be archived

Table A8.4 Details of data collected for CDM project activity

Data variable	Source of data	Data unit	Measured (m), calculated (c), or estimated (e)	Recording frequency	Proportion of data to be monitored	How will the data be archived? (electronic/ paper)	Comment

B.8.1.2 Data for monitoring of leakage (if applicable)

B.8.1.2.1 If applicable, please describe the data and information that will be collected to monitor the leakage of the proposed small-scale afforestation and reforestation CDM project activity

Table A8.5 Details of data collected for CDM project activity

Data variable	Source of data	Data unit	Measured (m), calculated (c), or estimated (e)	Recording frequency	Proportion of data to be monitored	How will the data be archived? (electronic/ paper)	Comment

B.8.2 Describe briefly the proposed QC and QA procedures that will be applied to monitor the actual GHG removals by sinks

B.8.3 Please describe briefly the operational and management structure(s) that the project operator will implement to monitor the actual GHG removals by sinks by the proposed small-scale afforestation and reforestation CDM project activity

B.9 Date of completion of the baseline study and the name of person (persons)/entity (entities) determining the baseline and the monitoring methodology

SECTION C: ESTIMATION OF EX-ANTE NET ANTHROPOGENIC GHG REMOVALS BY SINKS

C.1 Estimated baseline net GHG removals by sinks

C.2 Estimate of the actual net GHG removals by sinks

C.3 Estimated leakage

C.4 The sum of C.2 minus C.1 minus C.3 representing the net anthropogenic GHG removals by sinks of the proposed small-scale afforestation and reforestation CDM project activity

C.5 Table A8.6 provides values obtained when applying equations from the approved methodology

Table A8.6 Results obtained after applying equations from the approved methodology

Year	Estimation of baseline net GHG removals by sinks (tonnes of CO_2 equivalent)	Estimation of the actual net GHG removals by sinks (tonnes of CO_2 equivalent)	Estimation of leakage (tonnes of CO_2 equivalent)	Estimation of the net anthropogenic GHG removals by sinks (tonnes of CO_2 equivalent)
Year A				
Year B				
Year C				
Year …				
Total (tonnes of CO_2 equivalent)				

GHG – greenhouse gas

SECTION D: ENVIRONMENTAL IMPACTS OF THE PROPOSED SMALL-SCALE AFFORESTATION AND REFORESTATION CDM PROJECT ACTIVITY

D.1 Provide analysis of the environmental impacts, including trans-boundary impacts (if any)

D.2 If any negative impact is considered significant by the project participants or the host party, a statement that project participants have undertaken an environmental impact assessment, in accordance with the procedures required by the host party, including conclusions and all references to support documentation

D.3 Description of planned monitoring and remedial measures to address significant impacts referred to in section D.2 above

SECTION E: SOCIO-ECONOMIC IMPACTS OF THE PROPOSED SMALL-SCALE AFFORESTATION AND REFORESTATION CDM PROJECT ACTIVITY

E.1 Provide analysis of the socio-economic impacts, including trans-boundary impacts (if any)

E.2 If any negative impact is considered significant by the project participants or the host party, a statement that project participants have undertaken a socio-economic impact assessment, in accordance with the procedures required by the host party, including conclusions and all references to support documentation

E.3 Description of planned monitoring and remedial measures to address significant impacts referred to in section E.2 above

SECTION F: STAKEHOLDERS' COMMENTS

F.1 Brief description of how comments by local stakeholders have been invited and compiled

F.2 Summary of the comments received

F.3 Report on how due account was taken of any comments received

Annex 8.1: Contact Information on Participants in the Proposed Small-scale Afforestation and Reforestation CDM Project Activity

Table A8.7 Contact details of participants in the CDM project activity

Organization	
Street/PO Box	
Building	
City	
State/region	
Postfix/ZIP	
Country	
Telephone	
FAX	
E-mail	
URL	
Represented by	
Title	
Salutation	
Last Name	
Middle Name	
First Name	
Department	
Mobile	
Direct FAX	
Direct tel	
Personal e-mail	

Annex 8.2: Information Regarding Public Funding

Annex 8.3: Declaration on Low-income Communities

Please provide a written declaration that the proposed small-scale afforestation or reforestation project activity under the CDM is developed or implemented by low-income communities and individuals as determined by the host party.

Table A8.8 History of revisions in the project design document

Version	Date	Nature of revision
02	EB35, Annex 22 19 October 2007	• Sections A and B were restructured • Requirement to repeat equations has been removed from Section C • Sections D and E have been aligned with the requirements of the modalities and procedures
01	EB 23, Annex 16(a) and 16(b) 24 February 2006	Initial adoption

Appendix 9

List of Carbon Forest/ Agriculture Project and Their Comparison

The CDM Executive Board has approved certain rigorous baseline and monitoring methodologies that all CDM projects must use. A project can either submit a methodology for consideration or use the already approved methodologies. Until early 2011, five methodologies have been approved for agriculture, 11 for afforestation and reforestation, and six for agricultural residues/biogas. At present, only specific types of projects are allowed in the developing countries as per the rules for agriculture, forestry, and other land use (AFOLU) projects in CDM.

Table A9.1 Examples of agriculture and forestry carbon sequestration projects from the voluntary market

Types of projects	Examples	Who has done it?	Important details
Afforestation and reforestation projects (42%)	Uchindile–Mapanda Forest Projects in Tanzania: 10,814 ha of degraded land will be reforested and 7,565 ha put into conservation to protect local biodiversity.	Green Resources (www.greenresources.no) is the project developer. Community tree planting is promoted by giving away seedlings and providing necessary training in silviculture.	Generation of VERs over 99 years. Carbon revenues are locally reinvested and 10% is spent on community projects.
	Acacia Senegal Plantation Project, Mali. It will reforest over 17,700 ha of Acacia Senegal over a five-year period.	Deguessi Group, a private producer of agricultural products, is the project developer and works in partnership with local communities. The WB BioCarbon Fund buys the credits.	Intercropping is promoted. About 10,000 families receive social benefits; additional revenues generated by Arabic gum, grains, and forage and certified emission reduction.

Contd...

Table A9.1 *Contd...*

Types of projects	Examples	Who has done it?	Important details
REDD (28%)	REDD project in Sumatra/Indonesia, to protect the 750,000 ha Ulu Masen forest.	US investment bank Merrill Lynch, Aceh government, the British NGO Flora and Fauna International and Carbon Conservation. Funding comes from WB Multi-donor Fund's Aceh Environment and Forest project.	The project is expected to generate 100 million tonnes of VERs over 30 years.
	Noel Kempf Mercado Climate Action Project, Bolivia, for 30 years to protect 1.5 million ha of forests (www.noelkempff.com).	Government of Bolivia, the Friends of Nature Foundation, The Nature Conservancy, and three energy companies implement the $11 million project.	The aim is to stop logging activities and initiate alternative income programmes for communities. It has been verified in 2005 as the first forest emission reduction project.
Agricultural soil projects (16%)	Agricultural Carbon Project on 60,000 ha in Nyanza and Western Provinces, Kenya.	The project developer (Swedish Cooperative Center-Vi Agroforestry - SCC-ViA) promotes adoption of sustainable agricultural land management (cropland management and rehabilitation of degraded land). The WB BioCarbon Fund will purchase the credits.	Smallholder farmers and small-scale business entrepreneurs organized in common interest groups, primary level cooperatives, farmer groups and informal organizations. Funds will flow back to communities and farmers involved.
Plantations/ monoculture (13%)	Afforestation in the states of Odisha and Andhra Pradesh in India of 3,500 ha with tree plantations.	Promotion of plantations and agroforestry. Implementation is through a joint partnership between a cooperative society, a paper industry company and another company. The WB BioCarbon Fund will purchase the credits.	The plantations are established on bare and highly degraded land of small and marginal farmers on their private lands. Farmer cooperatives are supported to increase representation and negotiation power.

NGO – non-governmental organization; REDD – Reducing Emissions from Deforestation and forest Degradation; VER – voluntary or verified emission reduction

Note: The percentage of projects by category (in 2007) is indicated in brackets for each project type.

Source: Hamilton, Ryan, Emma, *et al.* (2008)

The Chicago Climate Exchange (CCX), an important trading system for GHGs, accepts those projects that generate emission reductions from forestry and rangeland activities occurring in the USA only.[1]

Some other examples of land-based carbon projects can be found in the inventory of the Forest Carbon Portal.[2]

[1] Details available at <www.chicagoclimatex.com>

[2] Details available at <www.forestcarbonportal.com>

Appendix 10

EXAMPLE OF AN INDIAN PDD WITH DETAILS*

CLEAN DEVELOPMENT MECHANISM PROJECT DESIGN DOCUMENT FORM FOR SMALL-SCALE AFFORESTATION AND REFORESTATION PROJECT ACTIVITIES (CDM-SSC-AR-PDD) (VERSION 02)

Section A: General Description of the Proposed Small-scale Afforestation and Reforestation CDM Project Activity

A.1 Title of the proposed small-scale afforestation and reforestation CDM project activity:

"Small-scale Cooperative Afforestation CDM Pilot Project Activity on Private Lands Affected by Shifting Sand Dunes in Sirsa, Haryana"

Version: 03

Date: 13/12/2008

A.2 Description of the proposed small-scale afforestation and reforestation CDM project activity:

The lands to be planted in the proposed small-scale afforestation and reforestation CDM project activity are located in the western belt of Haryana, which has its border with the state of Rajasthan at the north-eastern fringe of the Indian Thar Desert. The project area is affected by aeolian (wind-blown) sand and is the degraded part of croplands spread across these eight villages, comprising 369.87 ha belonging to 227 farmers; which is generally left fallow. Large areas of land are without any vegetation due to frequent dust storms of various intensities. These dust storms toss up large amount of sand, dust, and suspended particles into the air and pollute the ambient atmosphere. The report has found that the quality of drinking water has deteriorated and the water table

* Details available at <http://cdm.unfccc.int/UserManagement/FileStorage/DJISYMVG9RKHTCPU53X4BZA0NW8O2L>

has decreased over the years in this region. Many villages also have reportedly lost croplands due to shifting sands.[1] Impacted by limited precipitation (100–200 mm annually) and shifting sand dune, the cropping intensity on these degraded croplands is barely one crop every three years as against the normally two crops annually on the surrounding good croplands, as per participatory rural appraisal (PRA) findings. The cultivation and shifting sand dunes prevent the potential natural regeneration of forest in this area.

The objectives of the small-scale afforestation and reforestation CDM project activity proposed by Haryana CDM Variksh Kisan Samiti (Haryana CDM Tree Farmers Society), Ellenabad, Sirsa (hereafter known as the Society), are as follows:

- To earn carbon credits from the growing of trees to be planted, under the CDM provisions of the Kyoto Protocol.
- To help in the mitigation of global warming by planting trees for sequestration of atmospheric carbon dioxide.
- To improve the local environmental condition of soil through increasing the water-holding capacity of the lands, increasing the humus in soil, and also stabilizing sand dunes, by converting the marginal and degraded croplands into forested lands.
- To increase income, provide employment opportunities, and as a result alleviate the poverty of local communities.

To realize the aforementioned objectives, 369.87 ha of mixed forests will be established, using seven tree species: *Ailanthus excelsa, Acacia tortilis,* Eucalyptus hybrid, *Acacia nilotica, Dalbergia sissoo, Ziziphus mauritiana,* and *Prosopis cineraria*. The proposed small-scale afforestation and reforestation CDM project activity is a pilot project activity of its kind in Haryana. Both the project developer (Haryana Forest Department) and the local farmers (project participants) expect that the success of the proposed small-scale afforestation and reforestation CDM project activity will promote afforestation and reforestation CDM activities in lands of low agricultural productivity in Haryana and beyond the country. They are also of the view that it will contribute to poverty alleviation, biodiversity conservation, and prevention of soil erosion, thus contributing to sustainable development.

---

[1] Institute of Sustainable Development. Environment Impact Study in Haryana Community Forestry Project Villages, December 2007.

A.3 Project participants

Table A10.1 Project participants

Name of party involved[a] (host indicates a host party)	Private and/or public entity/ entities project participants (as applicable) (see a)	Indicate if the party involved wishes to be considered a project participant (Yes/No)
India	Haryana CDM Variksh Kisan Samiti, Ellenabad, Sirsa	No

CDM – Clean Development Mechanism

[a] At the time of making the CDM-SSC-AR-PDD public at the stage of validation, a party involved may or may not have provided its approval. At the time of requesting registration, the approval by the party/parties involved is required.

A.4 Description of location and boundary of the small-scale afforestation and reforestation CDM project activity:

The location and boundary of the proposed small-scale afforestation and reforestation CDM project activity is described in the following subsections.

A.4.1 Location of the proposed small-scale afforestation and reforestation CDM project activity

A.4.1.1 Host party/parties: India

A.4.1.2 Region/state/province: Haryana

A.4.1.3 City/town/community: The project area covers eight villages falling under three administrative blocks of Sirsa district, Haryana.

Table A10.2 Details of location and areas to be planted

State	District	Block	Village	Area to be planted (ha)	Coordinates	
					Latitude (N, degree)	*Longitude (E, degree)*
Haryana	Sirsa	Ellenabad	Neemla	45.89	29.252	74.351
			Dhani Sheranwali	73.27	29.241	74.363
			Bhuratwala	34.43	29.235	74.455
			Umedpura	42.65	29.244	74.495
			Poharkan	68.94	29.234	74.443
			Mallekan	26.85	29.245	74.565
Haryana	Sirsa	Sirsa	Madhosinghana	34.39	29.222	74.565
Haryana	Sirsa	Nathusari Chopta	Gudia Khera	43.46	29.223	74.573
Total				369.87		

A.4.2 Details of geographical location and project boundary, including information allowing the unique identification(s) of the proposed small-scale afforestation and reforestation CDM project activity:

The proposed small-scale afforestation and reforestation CDM project activity is located in the Sirsa district of Haryana state, which is 255 km from New Delhi. Sirsa lies between 29°14" and 30°0" north latitude and 74°29" and 75°18" east longitude, forming the western corner of Haryana, bordering the desert of Rajasthan in India (Figure A10.1). The lands to be planted in the proposed small-scale afforestation and reforestation CDM project activity are composed of 239 parcels of lands. The geographical axis, measured using global positioning system (GPS), at each corner of the boundary of the lands to be planted is shown in the maps in Figures A10.2 and A10.3. The hard copy of the maps will be available to the DOE for verification. Please also see GPS project boundary spreadsheet, attached with the PDD, showing geographical identification of each parcel of land.

Table A10.3 Details of lands to be planted in each village

Village	Area to be planted (ha)	Number of farmers	Number of land parcels
Bhuratwala	34.43	19	19
Umedpura	42.65	17	39
Poharkan	68.94	38	23
Mallekan	26.85	17	22
Madho Singhana	34.39	20	33
Gudia Khera	43.46	43	49
Total	369.87	227	270

A.5 Technical description of the small-scale afforestation and reforestation CDM project activity:

Technical issues of the proposed small-scale afforestation and reforestation CDM project activity are described in detail in the following subsections.

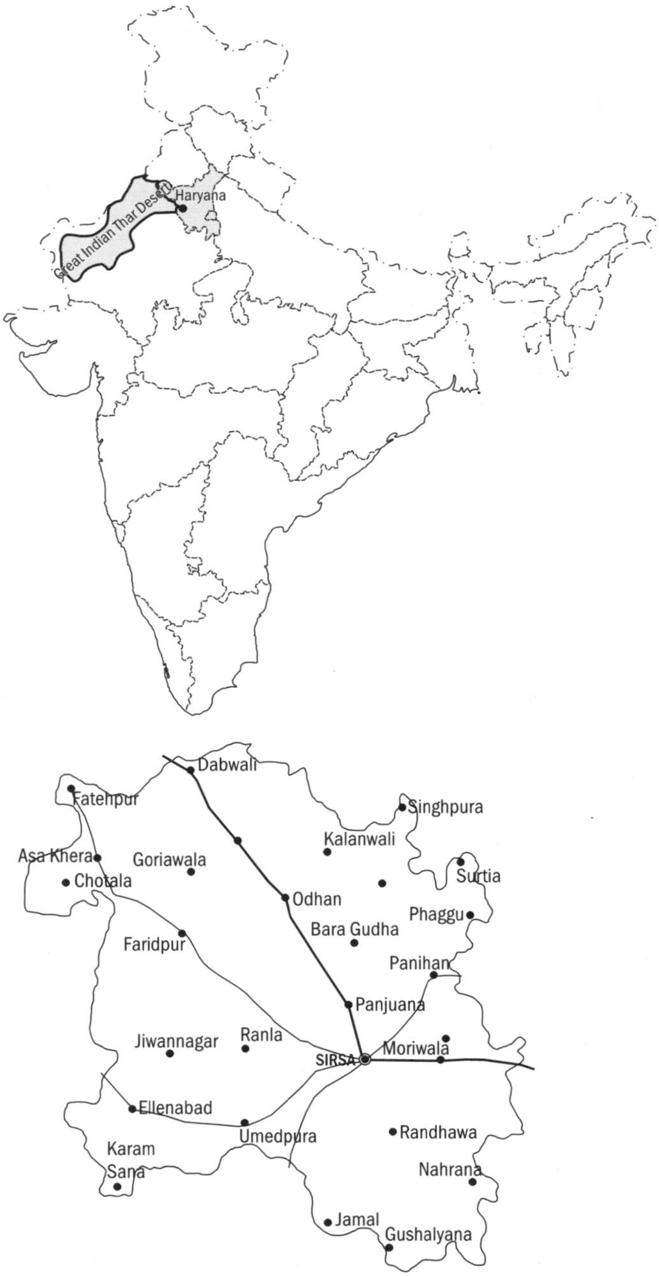

FIGURE A10.1 (a) Map of India showing the Great Indian Thar Desert and Haryana; project area is shown as a dot. (b) Map of Sirsa district of Haryana showing Ellenabad.

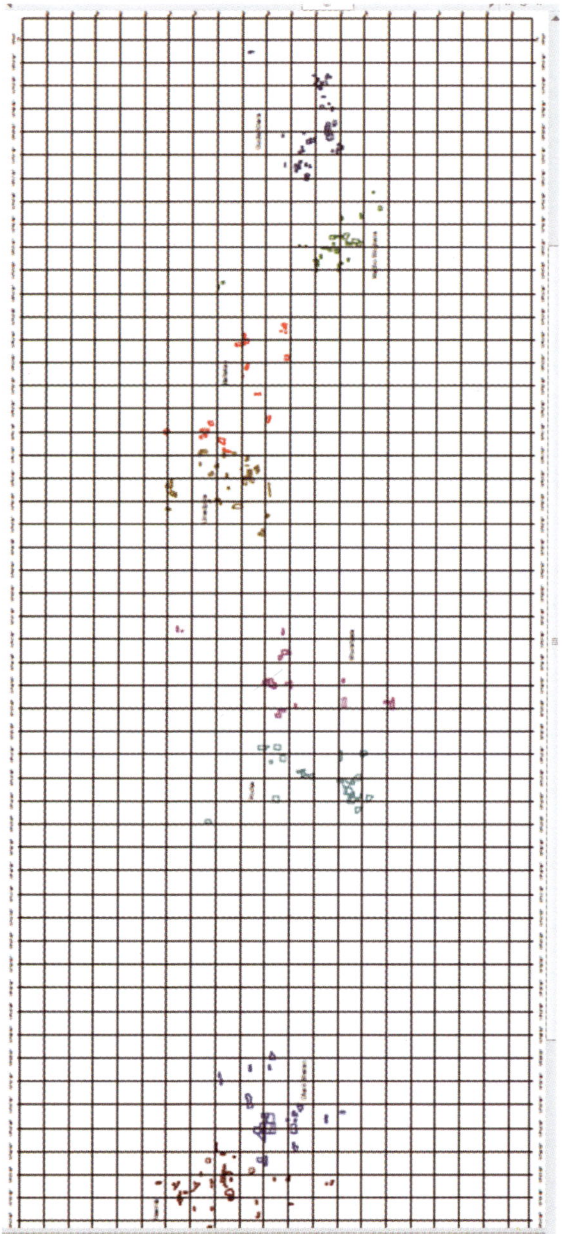

FIGURE A10.2 The project boundaries of all the areas included in the proposed project activity CDM Project Boundary in Sirsa District

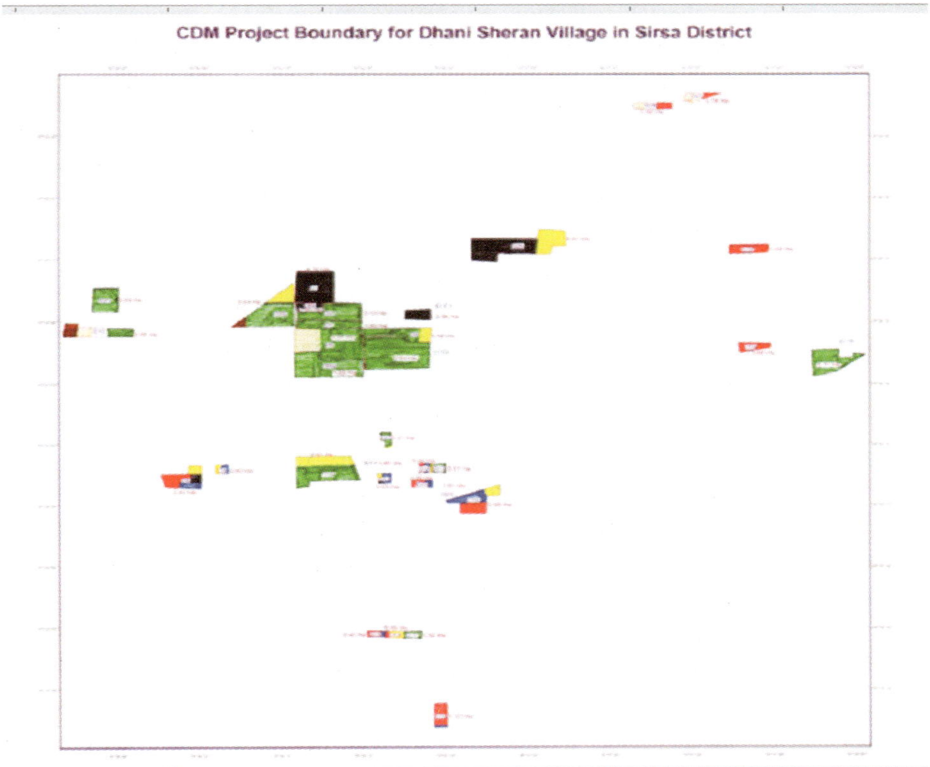

FIGURE A10.3 Project stratification map in Dhani Sheran village

A.5.1 Type(s) of small-scale afforestation and reforestation CDM project activity:

The project lands are marginal croplands that are cultivated once every three years in average, and are left fallow during other periods of time. Based on the Decision 6/CMP.1, titled "Simplified modalities and procedures for small-scale afforestation reforestation project activities under the clean development mechanism in the first commitment period of the Kyoto Protocol and measures to facilitate their implementation," the proposed small-scale afforestation and reforestation CDM project activity belongs to the type "cropland to forested land".

A.5.2 A concise description of the present environmental conditions of the area, which include information on climate, soils, main watershed, ecosystems, and the possible presence of rare or endangered species and their habitats:

For most part, the project area, located in the north-eastern fringe of the Great Indian Thar Desert, having been affected by mildly shifting sand dunes, is a degraded cropland. The region is affected by mild to severe sand storms blowing up volumes of dust and sand into the atmosphere and polluting the surrounding areas. The climate is arid, characterized by dryness and extremes of temperature. The mean daily maximum temperature during May and June, which is the hottest period, varies from 42°C to 47°C and winter temperature ranges between 2°C and 20°C. Precipitation is very low with average annual rainfall ranging between 150 mm and 200 mm. All these factors contribute to the occurrence of droughts in the area in summer and in the months of October and November. Frost is also common, usually occurring during the months of January and February. The area is also vulnerable to high velocity wind storms during May and June. All this has resulted in sparse and thin natural vegetation and no natural regeneration of trees in this area. Flood occurrences are not common in the area. So the drainage of water is badly developed and maintained, and when it rains, water logging has been reported in some villages, leading to salinity in soil and poor quality of potable water.

Desert or sandy soil is dominant in the area, which is wind-blown and light in colour. There is no perennial river in the area, and Ghaggar is the only river flowing from north to west through the central part of the Sirsa district. During the rainy season, its water is diverted into the southern canal flowing through these villages. Apart from this natural source of water supply, which is available only during the rains, there is a network of canals spread over these villages originating from the Bhakra Canal. These canals do not pass through the project area.

The old natural forests in these areas vanished in the 1950s due to the sudden increase in population on account of the influx of refugees in the aftermath of partition. It has emerged during the PRA that the migrants from Pakistan, who had temporarily settled in these villages, had cleared the vegetation for meeting their fuel needs as also for selling for earning livelihood. Subsequently, agriculture has been practised on most of these lands and presently, the proposed project lands have the status of degraded and degrading croplands, affected by shifting sand dunes, with a few scattered trees. The natural trees species available in the area are Jand (*Prosopis cineraria*), Beri (*Zizyphus mauritiana*), Jaal (*Salvadora oleoides*), Reru (*Acacia lucophloea*), Kair (*Capparis decidua*), Kikar (*Acacia nilotica*), and Pipal (*Ficus religiosa*) (also see Section C.1.b). None of these naturally occurring trees are rare or endangered according to the Botanical Survey of India.[2]

The wild animals that may be occasionally found in the project area include blackbuck (*Antelope cervicapra*), chinkara (*Gazella gazella*), desert cat (*Felis libyea*), desert fox (*Vulpus bucapus*), monitor lizard (*Varanus grizeus*), pea fowl (*Pavo cristatus*), jackal (*Carris aureus*), mongoose (*Herpester* sp.), and blue bull (*Boselaphus tragocamelus*). The major habitats of these wild animals are the sandy arid and semi-arid regions of Rajasthan, Madhya Pradesh, and Gujarat, and these animals occasionally wander into the fringe desert areas of Haryana, where the project area is also situated. The project area of 369.87 ha is an almost insignificant portion of the total habitat of these wild animals. Since the project area is scattered over eight villages, there is no habitat fragmentation or loss due to the project. There is no nature reserve in the vicinity of the project area. These lands currently have very low biodiversity.

---

[2] Details available at <http://envfor.nic.in/bsi/research.html>

A.5.3 Species and varieties selected:

Tree species to be used for planting have been determined by interviewing the farmers and taking into consideration carbon sequestration rates, biodiversity enhancement, soil and climate conditions, and the value of associated forest products. The chosen species include the following:

- *Ailanthus excelsa* (ullu neem)
- *Acacia tortilis* (Israeli kikar)
- Eucalyptus hybrid (safeda)
- *Acacia nilotica* (desi kikar)
- *Dalbergia sissoo* (shisham)
- *Zizyphus mauritiana* (desi ber)
- *Prosopis cineraria* (jand)

Of the above tree species, Eucalyptus hybrid (safeda) is an exotic species, and the rest are native to India, although not all occur naturally on the project lands. The exotic species has been selected because of high demand from farmers. Eucalyptus hybrid has been planted in this area for more than 50 years, and there are no reports of this species having become invasive.

A.5.4 Technology to be employed by the proposed small-scale afforestation and reforestation CDM project activity:

The technology to be employed under this pilot project is afforestation through direct planting of chosen species. Environment friendly technologies like the use of vermicompost, mulching, and spot irrigation will be used. National and international forestry experts will also be consulted for this purpose, but there will be no transfer of technology from an Annex 1 country to the host country.

To conserve the soil carbon stock and minimize the GHG emissions from the soil, zero tillage will be practised. Existing vegetation will not be removed, and small pits of size 45 cm × 45 cm × 45 cm will be dug. In the first two years, castor and other such crops will be grown by dibbling seeds without tilling the land, thus ensuring that the land over which soil disturbance has been caused, both due to tree planting and castor growing, will be less than 3% of the total project area. The planting material will be provided by the Haryana Forest Department from their nurseries established in the project villages.

The saplings will be raised in plastic containers, containing an appropriate mixture of soil and farmyard manure/vermicompost and humus to ensure healthy and robust initial growth. The species are divided into two categories: short-rotation species and medium-rotation species in the ratio of 4:6. This proportion will ensure regular flow of financial benefit to the farmers.

***Ailunthus excelsa* Seedlings Raised in Nursery**

A spacing of 4 m × 2.5 m will be maintained in the plantation (1000 trees per hectare) and replanting will be taken to fill up casualties. Planting will be done in blocks to maximize economic benefits. The harvesting of trees shall be done at the intervals of 10 and 20 years.

The design of plantation differs from one parcel of land to other. It was selected by the individual farmers according to their choice of species at the time of committing the lands for the proposed small-scale afforestation and reforestation CDM project activity. Please refer Table A10.4 for the details of parcel of lands and choice of species by farmers.

In the project area, short-rotation (10 years) species (40%) and medium-rotation (20 years) species (60%) will be planted. The species have been selected by the project participants under the guidance of the Haryana Forest Department, taking into account their carbon sequestration potential, biodiversity enhancement, soil and climate conditions, and the value of associated forest products, whereas rotation fixing has been largely a professional job by the forest department. The tree species selected including the two categories, that is, short rotation (10 years) and medium rotation (20 years) are given in Table A10.4.

Table A10.4 Tree species to be planted, number of trees, and rotation

Species	Number of trees in total area	Area (ha)	Rotation
Ailanthus excelsa (ullu neem)	26,300	57.86	Short rotation (10 years)
Acacia tortilis (Israeli kikar)	57,860	61.65	
Eucalyptus hybrid (safeda)	61,650	26.30	
Acacia nilotica (desi kikar)	53,650	60.75	
Dalbergia sissoo (sisham)	60,750	53.65	
Ziziphus mauritiana (desi ber)	74,200	35.46	Medium rotation (20 years)
Prosopis cineraria (jand)	35,460	74.20	
Total	369,870	369.87	

A.5.5 Transfer of technology/know-how, if applicable:

No technology will be transferred to the host country (India). The project is unilateral and no upfront CDM benefit has been sought for the farmers.

A.5.6 Proposed measures to be implemented to minimize potential leakage as applicable:

It has been ensured that the farmers contribute only small portions of their lands that are degraded and degrading and are unproductive and that they do not rely on these lands for their livelihood, so that the leakages due to the displacement of activities are unlikely to occur.

Although the simplified baseline and monitoring methodologies applied in the proposed small-scale afforestation and reforestation CDM project activity do not require emissions from transportation to be accounted under leakage, to minimize potential leakages bullock carts would be used to transport materials to and from the project area as much as possible.

A.6 A description of legal title to the land, current land tenure and land use, and rights to tCERs/lCERs issued:

The proposed project lands are owned by farmers. Current land use is cropland, but owing to the degraded feature of the lands, there is, on an average, only one crop every three years following good monsoon rains. All the timber, non-wood products, and carbon credits produced by the proposed small-scale afforestation and reforestation CDM project activity will be owned by the farmers who have the legal title to the lands.

In India there is no specific existing regulation on the rights to carbon credits. All benefits accruing from the project, including the CERs, belong to the owner of the land.

A.7 Assessment of the eligibility of land:

The Indian government defines forests as lands having trees with:

- a minimum area of 0.05 ha;
- a minimum tree crown cover of 15%; and
- trees of, or with potential to reach, the height of minimum 2 m.

The land eligibility is demonstrated using "procedures to define the eligibility of lands for afforestation and reforestation project activities" (EB 35 report Annex 18).

- At the moment when the project starts, the land is not a forest, which has been demonstrated by field survey and PRA. This has showed that the lands to be planted in the proposed small-scale afforestation and reforestation CDM project activity are currently degraded and degrading croplands affected by sand dunes and have very low productivity. Although some parcels of the project lands have existing mature trees, these trees are too sparse for any parcel to meet the definition of a forest. There are 492 trees in the project area (369.87 ha), and the crown cover constituted by these trees is about 0.75%.
- The activity is an eligible CDM afforestation project activity, which has been demonstrated through interviews with the local farmers/communities on land use/cover history and important events that have impacted the land use/cover changes. This has shown that the lands to be planted in the proposed small-scale afforestation and reforestation CDM project activity have been non-forest lands since the 1950s and hence satisfy the criteria of being non-forested since the last 50 years. See Sections E.1 and F.2 for details.

A.8 Approach for addressing non-permanence:

Please select between:

- Issuance of tCERs
- Issuance of lCERs

A.9 Duration of the proposed small-scale afforestation and reforestation CDM project activity/crediting period:

The crediting period for this project will be 20 years (with a choice of renewal twice for 20 years each), with verification occurring at intervals of every five years, followed by the issue of tCERs.

A.9.1 Starting date of the proposed small-scale afforestation and reforestation CDM project activity and of the (first) crediting period, including a justification:

The proposed small-scale afforestation and reforestation CDM project activity and its crediting period would start from 01/07/2008, for which the Haryana Forest Department would start establishing nurseries from December to January, 2007/08.

The actual plantation work would start from July 2008, because during this time, monsoon rains would increase the soil moisture content and would ensure survival of the saplings.

A.9.2 Expected operational lifetime of the proposed small-scale afforestation and reforestation CDM project activity:

The expected operational lifetime is 60 years.

A.9.3 Choice of crediting period and related information:

Please select one of the following:

- Renewable crediting period ✓
- Fixed crediting period ☐

A.9.3.1 Duration of the first crediting period (in years and months), if a renewable crediting period is selected:

The duration is 20 years 00 months.

A.9.3.2 Duration of the fixed crediting period (in years and months), if selected:

Not available

A.10 Estimated amount of net anthropogenic GHG removals by sinks over the chosen crediting period:

The net anthropogenic GHG removals by the sinks as a result of the proposed small-scale afforestation and reforestation CDM project activity are anticipated to be 231,920 tonnes of CO_2 equivalent during the crediting period (from 2008 to 2027). Table A10.5 gives the details.

Table A10.5 Ex-ante estimation of net anthropogenic GHG removals by sinks

Year	Annual estimation of net anthropogenic GHG removals by sinks in tonnes of CO_2 equivalent
2009	1,763
2010	1,784
2011	1,784
2012	1,784
2013	9,411
2014	9,411
2015	9,411
2016	9,411
2017	9,411
2018	−16,153
2019	10,852
2020	10,852
2021	10,852
2022	10,852
2023	29,742
2024	29,742
2025	29,742
2026	29,742
2027	29,742
Total estimated net anthropogenic GHG removals by sinks (tonnes of CO_2 equivalent)	231,920
Total number of crediting years	20
Annual average over the crediting period of estimated net anthropogenic GHG removals by sinks (tonnes of CO_2 equivalent)	11,596

GHG – greenhouse gas
Note: Minus sign indicates net emissions

A.11 Public funding of the proposed small-scale afforestation and reforestation CDM project activity:

There will be no public funding in the proposed small-scale afforestation and reforestation CDM project activity that will result in the diversion of official development assistance and financial obligations of any parties under the United Nations Framework Convention on Climate Change.

A.12 Confirmation that the small-scale afforestation and reforestation CDM project activity is not a de-bundled component of a larger project activity:

There is no registered small-scale afforestation and reforestation CDM project activity and no application to register another small-scale CDM project activity that conforms to the following criteria for determining the occurrence of de-bundling:

- Same project participants registered within the previous two years.
- Project boundary is within 1 km of the project boundary of the proposed small-scale afforestation and reforestation CDM activity at the closest point.

Therefore, the proposed small-scale afforestation and reforestation CDM project activity is not a de-bundled component of a larger project activity.

Section B: Application of a Baseline and Monitoring Methodology

B.1 Title and reference of the approved baseline and monitoring methodology applied to the proposed small-scale afforestation and reforestation CDM project activity:

Simplified baseline and monitoring methodologies for small-scale afforestation and reforestation project activities under the CDM implemented on grasslands or croplands (AR-AMS0001/Version 4.1)

B.2 Justification of the applicability of the baseline and monitoring methodology to the proposed small-scale afforestation and reforestation CDM project activity:

The applicability conditions laid down in the simplified baseline and monitoring methodologies for small-scale afforestation and reforestation project activities under the CDM implemented on grasslands or croplands (AR-AMS0001/Version 4.1) are as follows:

- Project activities are implemented on grasslands or croplands.
- Project activities are implemented on lands where the area of the cropland within the project boundary displaced due to the project activity is less than 50% of the total project area.

- Project activities are implemented on lands where the number of displaced grazing animals is less than 50% of the average grazing capacity of the project area.
- Project activities are implemented on lands where less than or equal to 10% of the total surface project area is disturbed as result of soil preparation for planting.

These applicability conditions are fulfilled as given below.

- The project area is located in the western belt of Haryana at the north-eastern fringe of the Great Indian Thar Desert. The area is severely affected by shifting wind-blown sands. The project area comprises 369.87 ha of highly degraded and degrading croplands beyond the reach of irrigation, which is generally left fallow, belonging to 227 farmers spread across eight villages. The cropping intensity on these degraded croplands is barely one crop in three years as against the norm of two crops per year on the surrounding good croplands.
- The farmers have contributed only a small portion of their lands, which are degraded and degrading, and they do not rely on these lands for their livelihood since these lands are unproductive. An agriculture crop is grown only once at an average interval of three years, and that too only in one cropping season after the monsoon rains, whereas two crops per year is the norm in the other irrigated agricultural lands in these eight and surrounding villages. The farmers have other productive lands on which they depend for their livelihood.

Table A10.6 Details of land holding of the farmers in each village

Village	Number of farmers	Total land holding (ha) of the project farmers	Land provided in the project (ha)
Neemla	43	205.0	45.89
Dhani Sherawali	30	118.4	73.27
Poharkan	19	207.0	68.94
Bhuratwala	17	139.3	34.43
Umedpura	38	166.0	42.65
Mallekan	17	72.4	26.85
Madho Singhana	20	156.6	34.39
Gudia Khera	43	201.0	43.46
Total	227	1,265.7	369.87

- Most cattle in the project villages are stall fed, most of the time with fodder grown on the irrigated agricultural lands outside the project lands. The only period they are allowed to graze is immediately after the crop harvest for a few days in the cropped lands belonging to the cattle owners. During April to May, the project villages are also visited by migratory cattle and herds of sheep and goat from the neighbouring state of Rajasthan. These migratory animals feed on grasses on government lands, roadsides, and also on the agricultural wastes left on the irrigated agricultural fields when allowed by the owner of the agricultural lands. The project lands do not support any grazing due to its degraded nature. Thus, there is no cattle displacement due to the project activity and hence the use of this simplified methodology is justified.
- A total of 1000 plants per ha will be planted in pits of the size 45 cm × 45 cm × 45 cm. In addition, in the first two years, up to 3000 castor plants will be sown in between the planting rows by dibbling seeds with not more than 10 cm^2 of soil being disturbed by each dibble. There will be no ploughing of land before the establishment of plantation. Thus, the planting activities will lead to soil disturbance over less than 3% of the surface area, and hence the use of this simplified methodology is justified.
- There is no single accepted definition of low-income communities by the Government of India, for the purpose of CDM projects. Therefore, the international norms have been followed according to which people below a daily income of $1 (around ₹50) fall under the poverty line. Most members of the society fall in this low-income category with an average annual income of just ₹6610 as has emerged in the PRA report (see E.1). Also the poor living conditions, with no toilet and running water facilities, in most households that have contributed their lands to the project justify their inclusion in the low-income category. In view of the aforementioned explanation, the use of a simplified methodology is justifiable.

B.3 Specification of the GHGs whose emissions will be part of the proposed small-scale CDM project activity:

The GHG that will be part of the project emissions is N_2O. These emissions are estimated to be negligible based on preliminarily estimation. The spreadsheet for calculation of emissions shall be made available to the DOE during verification.

B.4 Carbon pools selected:

Based on the simplified baseline and monitoring methodology applied by the proposed small-scale afforestation and reforestation CDM project activity, the above-ground and below-ground biomass (living biomass) are the only carbon pools to be considered (Table A10.7).

Table A10.7 Chosen carbon pools

Carbon pools	Selected (answer with yes or no)
Above ground	Yes
Below ground	Yes
Dead wood	No
Litter	No
Soil organic carbon	No

B.5 Description of strata applied for ex-ante estimations:

No stratification has been done for the ex-ante baseline estimation since the climate, landform, soil conditions influencing carbon stocks, in accordance with Section 4.3.3.2 of *IPCC Good Practice Guidance for Land use, Land-use Change, and Forestry* (IPCC 2003), are largely similar over the entire project area. Also the method adopted for baseline estimation is total count rather than sampling.

For the ex-ante estimation of the carbon stocks in the project area, the climate, landform, soil conditions, and afforestation activity, including planting density and age, being the same throughout the project area, the project lands have been stratified into seven project strata based on species alone. Each species represents one stratum.

B.6 Application of baseline methodology to the proposed small-scale afforestation and reforestation CDM project activity:

The baseline approach, "existing or historical, as applicable, changes in carbon stock in the carbon pools within the project boundary", is the most appropriate approach for determining the baseline scenario. This is because of the degraded and degrading nature of the project lands caused by the shifting sand dunes, the poor condition of landowners, and the very high cost of raising and maintaining trees on such lands. The lands to be afforested, without the proposed small-scale afforestation and reforestation CDM project activity, would continue to remain in their present status. The efforts of the Haryana State Forest Department to encourage the farmers to promote tree planting under their social forestry, farm forestry, and community forestry programmes over the past 20 years, generally considered successful in achieving their objectives in the neighbouring lands, have not been successful in raising trees on the project lands on account of the exceptionally degraded nature of these lands. The baseline approach "the most likely land use would be continued current land use" also carries the same meanings in this context.

The baseline approach "changes in carbon stocks in the carbon pools within the project boundary from a land use that represents an economically attractive course of action" is not appropriate because there is no economically more attractive alternative to the current status of these lands unless CDM finance is available.

The carbon stock in the living biomass pools of woody perennials and grasslands is expected to decrease in the absence of the proposed small-scale afforestation and reforestation CDM project activity, and is conservatively assumed to be constant (see C.1a). However, due to the continuous growth of the pre-project trees in the baseline scenario, the baseline net GHG removal by sinks is calculated as the increase in carbon stock in living biomass of the pre-project living trees even though most of these living trees are mature (see C.1b). In the above case, the baseline carbon stocks in the carbon pools equal to the existing carbon stocks at the start of the proposed small-scale afforestation and reforestation CDM project activity.

B.7 Description of how the actual net GHG removals by sinks are increased above those that would have occurred in the absence of the registered small-scale afforestation and reforestation CDM project activity:

The condition of additionality has been fulfilled as no tree-planting activity with comparable sequestration capacity would have been taken up on the project lands in the absence of the project due to the non-viability of such an activity on account of the degraded and degrading nature of these lands. The Community Forestry Project of the Haryana Forest Department has been very active in this area targeting private lands in the past seven years, but the lands in question were not taken up for tree plantation under the project for the same reason. Also, since the lands in question are privately owned degraded croplands, the afforestation of these lands is not a legal requirement. Thus, the project lands would not have been taken up for tree planting in the baseline scenario. This has been established by barrier analysis presented as follows.

- **Investment barriers:** No credit mechanisms are in place for farmers to make long-term investment in plantation forestry on degraded and degrading lands by taking commercial loans from banks. Agriculture is the main income source in the project area, and with low productivity, the condition of most farmers borders on poverty. As a result, they are not able to afford the high plantation costs, particularly with the long gestation periods that forestry entails. The proposed small-scale afforestation and reforestation CDM project activity reduces the gestation period for economic returns through carbon credits and makes the project a more attractive economic proposition.

  The Haryana Forest Department has taken up the proposed small-scale afforestation and reforestation CDM project activity as a pilot project with a view to promote tree planting for climate change mitigation on the most unproductive lands. It is bearing all project investigation and preparation costs as well as nursery costs in the first phase and will also endeavour to help the project proponent raise money for raising plantation. This is being done only because it is a CDM project.
- **Technological barriers:** The local farmers do not have an easy access to either planting material or planting technologies as forestry is not their usual occupation. They get these facilities only through the Haryana Forest Department under its social and community forestry project. But owing to the lack of resources with the forest department, these projects have so far been confined to only better and more productive lands in

view of the very high costs of raising plantations in lands as unproductive as the proposed project lands. The Haryana Forest Department has been able to spare resources for raising trees on these lands only because of its commitment to promotion of CDM afforestation on such lands.

- **Barriers due to social conditions**: No organization of local communities is focused on tree planting, and individual farmers are unable to successfully invite investments in tree planting on such lands and to exploit commercial synergies with their other products and create new links to market on their own. This also prevents them from overcoming technological barriers mentioned earlier. Now the interest in CDM and the benefits that are likely to flow from a CDM project have led to the creation of the Haryana CDM Variksh Kisan Samiti, Ellenabad, Sirsa, specifically for this purpose under the patronage of the Haryana Forest Department. Thus an important barrier to tree planting on the project lands has been removed because of CDM.

The natural regeneration would not be possible and would not result in the conversion of the project lands to a forest in the absence of the proposed small-scale afforestation and reforestation CDM project activity due to the following reasons:

- Intermittent agricultural cultivation over these lands for a long time has resulted in complete loss of root stock of trees, thus preventing the possibility of restocking through regeneration from root suckers.
- There are no forests in the vicinity, thus preventing the possibilities of large-scale seed availability for germination. Where seeds do disperse into the edge of the land parcels or from pre-project existing trees, intermittent cultivation prevents the survival of any seedlings.

Carbon dioxide will be sequestered from the atmosphere through the growth of planted trees and stored in the above-ground and below-ground biomass of living trees. In the project scenario, 1000 trees per hectare will be planted, resulting in much higher carbon stock change in living biomass than in the baseline scenario. It can also be demonstrated that the increase in GHG emission is negligible. Therefore, the actual net GHG removals by sinks will be increased above those that would have occurred in the absence of the registered small-scale afforestation and reforestation CDM project activity.

Refer Section C for detailed methods and ex-ante estimation of baseline net removals by sinks and actual net GHG removals by sink.

B.8 Application of monitoring methodology and monitoring plan to the small-scale afforestation and reforestation CDM project activity:

a. Ex-post estimation of the baseline net GHG removals by sinks

In accordance with the decision 6/CMP.1, Appendix B, paragraph 6, no monitoring of the baseline is required for small-scale afforestation and reforestation CDM project activity. The baseline net GHG removals by sinks will be assumed to be those estimated in C.1.

b. Ex-post estimation of the actual net GHG removals by sinks

The project participants will determine any changes in carbon stocks via measuring and monitoring the project area that has been planted. The monitoring will focus on (1) the project boundaries and (2) the selected stratified sample plots. The stratification shall be based on the species planted, other things being almost uniform throughout the project lands (see B.5) and carbon sampling will take place within stratified project area. All sampling will be in accordance with the methods described in 4.3.3.4 of the *IPCC Good Practice Guidance for Land use, Land-use Change, and Forestry* (IPCC 2003). This monitoring plan will be used throughout the project area and the crediting period. If at a later stage, it is found that the carbon stocks in some areas differ significantly from those in the same strata elsewhere, these areas will be assessed as a separate stratum.

The project boundary will be monitored by monitoring of parcels using GPS. Any changes in project boundary will be accounted for in all calculations of the actual net GHG removals by sinks. The monitoring methodology uses permanent sample plots to monitor carbon stock changes in above-ground and below-ground biomass pools. To reach the targeted precision level of about ±10% of the mean at 95% confidence level in a cost-effective manner, the number of plots needed in each stratum (Table A10.8) has been determined following Equations (B.1) and (B.2) in the draft methodological tool "calculation of the number of sample plots for measurements within afforestation and reforestation CDM project activities" approved by Executive Board:[3]

[3] Details available at <http://cdm.unfccc.int/EB/Meetings/031/eb31_repan15.pdf>

$$n = \frac{\left(\sum_{i-1}^{L} N_1 {*} st_i\right)^2}{\left(\frac{N{\cdot}E}{z_{\alpha/2}}\right)^2 + \left(\sum_{i-1}^{L} N_1 {*} (St_i)^2\right)} \tag{B.1}$$

$$n_i = \frac{\sum_{i-1}^{L} N_1 \cdot St_i}{\left(\frac{N{\cdot}E}{z_{\alpha/2}}\right)^2 + \sum_{i-1}^{L} N_i \times (st_i)^2} \cdot N_i \cdot st_i \tag{B.2}$$

$$N = \frac{A}{AP}\, N_1 = \frac{A_i}{A_P}$$

$$E = Q{\cdot}p \tag{B.4}$$

Here

- n is the sample size (total number of sample plots required) in the project area;
- n_i is the sample size for stratum i;
- E is the allowable error of the estimated quantity Q;
- i is the project strata;
- L is the total number of strata (dimensionless);
- $1-\alpha$ is the probability that the estimate of the mean is within the error bound E;
- $z_{\alpha/2}$ is the value of the statistic z (embedded in excel as inverse of standard normal probability cumulative distribution), for example, $1-\alpha = 0.05$ (implying a 95% confidence level), $z_{\alpha/2} = 1.9599$;
- N_i is the maximum possible number of sample plots in stratum I;
- N is the maximum possible number of sample plots in the project area;
- St_i is the standard deviation for each stratum i (dimensionless, 30%);
- A is the total size of all strata, for example, the total project area (ha);
- A_i is the size of each stratum i (ha);
- AP is the sample plot size (constant for all strata) (0.04 ha);

- Q is the approximate average value of the estimated quantity (above-ground wood volume per hectare) (m^3/ha);
- *p* is the desired level of precision (10%) (dimensionless).

Table A10.8 Number of plots to be sampled for different strata

Strata	Defining feature of the strata (species)	Area (ha)	Number of sampling plots
S1	Eucalyptus hybrid	26.30	4
S2	*Ailanthus excelsa*	57.86	6
S3	*Acacia tortilis*	61.65	7
S4	*Dalbergia sissoo*	53.65	5
S5	*Acacia nilotica*	60.75	5
S6	*Prosopis cineraria*	74.20	3
S7	*Zizyphus mauritiana*	35.46	3
Total		369.87	32

Plots located by GPS ensure the measuring and monitoring consistently over time. To avoid subjective choice of plot locations (plot centres, plot reference points, movement of plot centres to more "convenient" positions), the permanent sampling plots shall be located systematically with a random start, which is considered good by *IPCC Good Practice Guidance for Land use, Land-use Change, and Forestry.* This can be accomplished with the help of a GPS in the field. The geographical position (GPS coordinate), administrative location, and stratum series number of each plots shall be recorded and archived. The size of plots will be 400 m^2. Also, it is to be ensured that the sampling plots are distributed as evenly as possible.

The carbon stocks, expressed in CO_2 equivalent, will be estimated through stratified random sampling procedures and using the following equations:

$$P_{(t)} = \sum_{i=1}^{L} (P_{A(t)i} + P_{B(t)i})^* A_i^*(44/12)$$

Here

- $P_{(t)}$ is the carbon stocks within the project boundary at time *t*, achieved by the proposed small-scale afforestation and reforestation CDM project activity (tonnes of CO_2 equivalent);

- $P_{A(t)i}$ is the carbon stocks in above-ground biomass of stratum i at time t, achieved by the proposed small-scale afforestation and reforestation CDM project activity during the monitoring interval (tonnes of carbon per hectare);
- $P_{B(t)i}$ is the carbon stocks in below-ground biomass of stratum i at time t, achieved by the proposed small-scale afforestation and reforestation CDM project activity during the monitoring interval (tonnes of carbon per hectare);
- A_i is the area of stratum i (ha);
- i is stratum number (I is the total number of strata).

For above-ground biomass, $P_{A(t)i}$ is calculated per stratum i as follows:

$$P_{A(t)i} = E_{(t)I} \times 0.5$$

where

- $P_{A(t)i}$ is carbon stocks in above-ground biomass at time t, achieved by the proposed small-scale afforestation and reforestation CDM project activity during the monitoring interval (tonnes of carbon per hectare);
- $E_{(t)i}$ is the estimate of above-ground biomass at time t, achieved by the proposed small-scale afforestation and reforestation CDM project activity (tonnes of dry matter per hectare);
- 0.5 is the carbon fraction of dry matter (tonnes of carbon/tonnes of dry matter).

The calculations shall be performed for each stratum. The estimate of above-ground biomass at time t achieved by the proposed small-scale afforestation and reforestation CDM project activity $E_{(t)}$ shall be estimated through the following steps:

- **Step 1**: Establish permanent sample plots and document their location in the first monitoring report.
- **Step 2**: Measure the diameter at breast height (DBH) and tree height in the sample plots.
- **Step 3**: Estimate the above-ground stand volume using the following local allometric equations:

- *Acacia* sp.
  $V = -0.00142 + 2.61911\ D - 0.54703\ D_{0.5}$
  (n = 151, R_2 = 0.95768); applicable for eastern Rajasthan
  (FSI 1996, page 13)
- Eucalyptus hybrid
  $V = 0.02894 - 0.89284\ D + 8.72416\ D_2$
  (n = 198, R_2 = 0.9892); applicable for drier parts of Karnataka
  (FSI 1996, page 93)
- Dalbergia sissoo
  $V = -0.3238 + 3.0077\ D$
  (n = 1146, R_2 = 0.9358); applicable for Haryana
  (FSI 1996, page 221)
- *Zizyphus* spp.
  $V = -0.002557 + 0.260114\ D_2H$
  (n = 142, R_2 = 0.93), General Volume Table applicable to *Zyzhyphus xylopara* for drier parts of Andhra Pradesh
  (FSI 1996, page 195)
- Ailanthus excelsa
  $V = -0.09362 + 9.93014\ D_2$
  (n = 23, R_2 = 0.96323); applicable to *Ailanthus grandis* in Arunachal Pradesh
  (FSI 1996, page 22)
- Prosopis cineraria
  $V = 0.00471 + 1.79326\ D_2$
  (n = 1060, R_2 = 0.9041), local volume table applicable for miscellaneous species in eastern parts of Rajasthan bordering Haryana
  (FSI 1996, page 201)

The research wing of the Haryana Forest Department and also the Forest Research Institute, Dehradun, are developing new volume tables, which are expected to be published well before the date of first monitoring. In case new allometric equations are available at the time of first monitoring, the same shall be used instead of the above approximations.

- **Step 4**: Use biomass expansion factors as follows:

$$E_{(t)i} = SV_{(t)i} * BEF * WD$$

where

- $E_{(t)i}$ is the estimate of above-ground biomass of stratum i at time t achieved by the proposed small-scale afforestation and reforestation CDM project activity (tonnes of dry matter per hectare);
- $SV_{(t)i}$ is the stem volume (m^3/ha);
- WD is the basic wood density (tonnes of dry mater per cubic matter);
- BEF is the biomass expansion factor (over bark) from stem to the total above-ground biomass (dimensionless).

Table A10.9 Parameters to be used for different species

Species	Wood density	Biomass expansion factor	Root to shoot ratio
Eucalyptus hybrid	0.64	3.4	0.27
Ailanthus excelsa	0.64	3.4	0.27
Acacia tortilis	0.76	3.4	0.27
Dalbergia sissoo	0.64	3.4	0.27
Acacia nilotica	0.76	3.4	0.27
Prosopis cineraria	0.64	3.4	0.27
Ziziphus mauritiana	0.76	3.4	0.27
Reference	Table 3A.1.9-2 of *IPCC Good Practice Guidance for Land use, Land-use Change, and Forestry*	Table 3A.1.10 of *IPCC Good Practice Guidance for Land use, Land-use Change, and Forestry*	Table 3A.1.8 of *IPCC Good Practice Guidance for Land use, Land-use Change, and Forestry*

BEF – biomass expansion factor; IPCC – Intergovernmental Panel on Climate Change
Source: IPCC (2003)

Carbon stocks in below-ground biomass at time t, achieved by the proposed small-scale afforestation and reforestation CDM project activity during the monitoring interval $P_{B(t)}$ shall be estimated for each stratum i as follows:

$P_{B(t)i} = E_{(t)i} \times R \times 0.5$
where

- $P_{B(t)i}$ is carbon stocks in below-ground biomass at time t achieved by the proposed small-scale afforestation and reforestation CDM project activity during the monitoring interval (tonnes of carbon per hectare);
- $E_{(t)i}$ is the estimate of above-ground biomass of stratum i at time t achieved by the proposed small-scale afforestation and reforestation CDM project activity (tonnes of dry matter per hectare);
- R is the root to shoot ratio (dimensionless);
- 0.5 is the carbon fraction of dry matter (tonnes of carbon/tonnes of dry matter).

B.8.1 Data to be monitored: Monitoring of the actual net GHG removals by sinks and leakage.

B.8.1.1 Actual net GHG removals by sinks data:

B.8.1.1.1 Data to be collected or used to monitor the verifiable changes in carbon stock in the carbon pools within the project boundary resulting from the proposed small-scale afforestation and reforestation CDM project activity, and how this data will be archived:

Table A10.10 Details of data collected to monitor verifiable changes

Data variable	Source of data	Data unit	Measured (m), calculated (c), or estimated (e)	Recording frequency	Proportion of data to be monitored	How will the data be archived? (electronic/ paper)	Comment
Location of the areas where the project activity has been implemented	Field survey using GPS	Latitude and longitude	Measured	4	100%	Electronic, paper, photos	GPS will be used for field survey
Size of the areas where the project activity has been implemented for each type of strata (A_i)	Field survey using GPS	ha	Measured	4	100%	Electronic, paper, photos	GPS will be used for field survey

Contd...

Table A10.10 *Contd...*

Data variable	Source of data	Data unit	Measured (m), calculated (c), or estimated (e)	Recording frequency	Proportion of data to be monitored	How will the data be archived? (electronic/ paper)	Comment
Land-use rights	Land records with the revenue department	ha	Estimated	4	100%	Paper	Land registration certificates from the revenue department may be verified
Location of the permanent sample plots	Project maps and project design	Latitude and longitude	Defined	4	100%	Electronic, paper	Plot location is registered with a GPS and marked on the map
Diameter of tree at breast height (1.3 m)	Permanent sample plots	cm	Measured	4	Each tree in the sample plot	Electronic, paper	Measure diameter at breast height for each tree that falls within the sample plot and applies to size limits
Height of tree	Permanent sample plot	M	Measured	4	Each tree in the sample plot	Electronic, paper	Measure height (H) for each tree that falls within the sample plot and applies to size limits
Basic wood density	IPCC default value	Tonnes of dry matter per m^3	Estimated	Once	Each tree in the sample plot	Electronic, paper	
Biomass expansion factor	IPCC default value	Dimension-less	Estimated	Once	Each tree in the sample plot	Electronic, paper	
Root–shoot ratio	IPCC default value	Dimension-less	Estimated	Once	Each tree in the sample plot	Electronic, paper	
Total CO_2	Project activities	mg	Calculated	4	All project data	Electronic	Based on the data collected from all plots and carbon pools

GPS – geographic information system; IPCC – Intergovernmental Panel on Climate Change

B.8.1.2 Data for monitoring of leakage (if applicable):

Although leakages are unlikely to occur, they will still be monitored through participatory method through the assessment of displacement of people, agriculture production, fuelwood, timber, and grazing.

B.8.1.2.1 If applicable, please describe the data and information that will be collected to monitor the leakage of the proposed small-scale afforestation and reforestation CDM project activity.

Table A10.11 Details of data collected to monitor leakage

Data variable	Source of data	Data unit	Measured (m), calculated (c), or estimated (e)	Recording frequency	Proportion of data to be monitored	How will the data be archived? (electronic/ paper)	Comment
Percentage of families displaced due to project implementation	PRA	Number of families or households	Estimated	4	100%	Electronic, paper	
Percentage of total grain production displaced due to project activity	PRA	Quantity	Estimated	4	100%	Electronic, paper	
Percentage of total fuelwood procurement displaced due to project activity	PRA	Quantity	Estimated	4	100%	Electonic, paper	
Percentage of total timber procurement displaced due to project activity	PRA	Quantity	Estimated	4	100%	Electronic, paper	
The time-average number of grazing animals per hectare within the project boundary that is displaced due to the project activity divided by the average grazing capacity of land for the area, expressed as percentage	PRA	Grazing animal days/ha/ year	Estimated	4	100%	Electronic	

PRA – participatory rural appraisal

B.8.2 Describe briefly the proposed quality control (QC) and quality assurance (QA) procedures that will be applied to monitor actual GHG removals by sinks:

A rigid QA and QC method involving standard operating procedures to ensure the (1) reliability of collection of field measurements, (2) verification of the methods used to collect field data, (3) verification of data entry and analysis techniques, (4) verification of data maintenance and archiving, and (5) upgrading of electronic data with technological changes shall be followed as discussed below.

QA and QC for field measurements: Collecting reliable field measurements is an important step in the QA plan. Those responsible for the carbon measurement work will be fully trained in all aspects of the field data collection and data analyses, and standard operating procedures, described below, will be followed rigidly to ensure accurate measurement and re-measurement. The SOPs include auditing procedures. The first type of audit, a 'hot check', consisting of the project leader observing field crew members during data collection to ensure field measurements SOPs shall be followed to correct any technique errors. The second type of audit, a 'blind check', to quantify measurement errors, shall consist of a complete re-measurement of 10% of plots by verifying teams highly experienced in forest measurement and attentive to detail. After measurement a comparison will be made with the original data and discrepancies re-verified. Any errors found will be corrected and recorded. Any errors discovered will be expressed as a percentage of all plots that have been rechecked to provide an estimate of the measurement error.

For all the Verified Plots

QA/QC for data entry: To produce reliable carbon estimates, proper entry of data in the data analysis spreadsheets is required. Steps will be taken to ensure that errors are minimized. Results will be reviewed and double-checked for typing errors. If there are any problems with monitoring plot data that cannot be solved, the plot should not be used in the analysis.

QA/QC for data archiving: Due to the long-term length of project activities, data storage and maintenance is very important. SOPs for data archiving include storage of original copies of all data, including field measurements, laboratory measurements, and Geographic Information System products, and copies of measuring and monitoring reports, in a secure off-site location and provision of copies of all data to all project participants. Copies of all data will also be stored on paper in a separate remote location. Procedures also include updating storage onto new data storage technologies, both hardware and software.

B.8.3 Please describe briefly the operational and management structure(s) that the project operator will implement to monitor actual GHG removals by sinks by the proposed small-scale afforestation and reforestation CDM project activity:

The proposed small-scale afforestation and reforestation project activity will be implemented by the society, under the guidance of the Haryana Forest Department. The department shall provide all needed technical help, including training, to the persons selected by the society and will facilitate consultation with other experts wherever such consultations become necessary.

The Haryana Forest Department shall also coordinate the measuring and monitoring of the actual GHG removals by sinks and any leakage generated by the proposed small-scale afforestation and reforestation CDM project activities. Any activity data and monitoring and measuring data will be reported to and archived in the department headquarter office at Panchkula, which shall act as the State Carbon Sequestration Office.

The Conservator of Forests in Hisar, under the guidance and coordination of the Principal Chief Conservator of Forests of the Haryana Forest Department acting as the State Carbon Sequestration Office, will provide instruction for afforestation and forest management, and will supervise the implementation of the proposed small-scale afforestation and reforestation CDM project activity, and collect specific activity data.

The Principal Chief Conservator of Forests shall arrange for technical consultation and training in the measuring and monitoring of the actual GHG removal by sinks and leakage generated by the proposed small-scale afforestation and reforestation CDM project activity, and shall be responsible for helping the society draft the monitoring report. An expert team shall be established by the Principal Chief Conservator for addressing any technical issues arising out of checking and verification of measured and monitored data.

B.9 Date of completion of the baseline study and the name of person (persons)/entity (entities) determining the baseline and the monitoring methodology:

Date of completion of the baseline study: 10/01/2008

Name of person (persons)/entity (entities) determining the baseline and the monitoring methodology:

- Xiaoquan Zhang (xiaoquan@caf.ac.cn), Institute of Forestry Ecology, Environment and Protection, the Chinese Academy of Forestry
- Institute of Global Warming & Ecological Studies, Noida, India
- Promode Kant (promode.kant@gmail.com)
- Keshav C Das (kcdas@amity.edu)
- Debojyoti Chakraborty (dchakraborti@amity.edu)
- Ashwin A. S. (as.ashwin@gmail.com)
- Sohini Trehan (sohini_trehan@rediffmail.com)
- Swati Singh (swati123singh@gmail.com)
- Kundan Burnwal (k.burnwal@gmail.com)
- M.A. Khalid (makhalid@amity.edu)
- Haryana Forest Department
- S.K. Dhar (dharsk_ifs@yahoo.co.in)
- Jeet Ram (hcfp2003@yahoo.com)
- Goran Jonsson (hcfp2003@yahoo.com)
- D.R. Ramesh Singh (drramesh2004@yahoo.com)

- V.S. Tanwar (vstanwar@rediffmail.com, vst1987@gmail.com)
- S.S. Sheoran (vst1987@gmail.com)
- Institute of Climate Change and Ecology, New Delhi

  With effect from 01/05/2008, the revision of PDD has been undertaken by Promode Kant, Ashwin A. S., and Kundan Burnwal on behalf of the Institute of Climate Change and Ecology, New Delhi.

SECTION C: ESTIMATION OF EX-ANTE NET ANTHROPOGENIC GHG REMOVALS BY SINKS

C.1 Estimated baseline net GHG removals by sinks:

C.1.1 Pre-project carbon stock

C.1.1.1 Non-tree vegetation

The woody perennials (shrubs and herbaceous plants) existing at the project area before the start of the proposed small-scale afforestation and reforestation CDM project activity were counted and measured. A total of 2093 woody perennials exist in the project area with an average height of 0.7 m (average number of woody perennials per hectare is 5.66).

Since the area is shifting sand-dune-affected degraded and degrading land, there is little growth of grasses as evident in the field visits. The total biomass of the woody perennials and grass was calculated by taking 30 kg of dry matter per hectare as per the expert judgement. The total biomass of woody perennials and herbaceous plants comes to 11 tonnes of dry matter. The carbon factor is 50% of the dry weight. Hence, the carbon stock of the woody perennials is 5.5 tonnes.

C.1.1.2 Pre-project living tree

There are 492 trees pre-existing in the area specified for the proposed small-scale afforestation and reforestation CDM project activity. The DBH, tree height, crown diameter, and age of all pre-project living trees have been measured. Volume has been estimated from the local volume tables[4] provided by the Haryana Forest Department. The stem volume of trees was then converted into carbon stock in above-ground and below-ground biomass through wood density (D), biomass expansion factor (BEF), root–shoot ratio (R), and carbon fraction (CF) using the following equations:

$$C_{AB} = V \times D \times BEF \times CF$$

$$C_{BB} = C_{AB} \times R$$

where

- C_{AB} is the carbon stock in above-ground biomass (tonnes of carbon per tree)
- C_{BB} is the carbon stock in below-ground biomass (tonnes of carbon per tree) (tonnes of carbon per tree)
- V is the stem volume (m^3 per tree)
- D is the wood density (tonnes of dry matter per m^{-3})
- BEF is the biomass expansion factor from stem biomass to above-ground biomass (dimensionless)
- CF is carbon fraction (tonnes of carbon/tonne of dry matter; IPCC default = 0.5)
- R is the root–shoot ratio (dimensionless)

D, BEF, and *R* are from Table A10.9. IPCC default value (0.5) is used for the carbon fraction. The carbon stock in living biomass of the pre-project trees was then estimated and provided in Table A10.12.

C.2 Estimation of baseline net GHG removals by sinks:

Due to the impact of shifting sand dunes and agricultural cultivation, lands are degraded and degrading. The carbon stocks, both in the living biomass pool of woody perennials and in the below-ground

[4] Letter no. 280, Indira Gandhi Canal Project, Forest Department, Government of Rajasthan

biomass of grasslands, are expected to decrease in the absence of the proposed small-scale afforestation and reforestation CDM project activity. Therefore, based on paragraph 6(a) and (b) of the methodology applied, the baseline carbon stocks in the living biomass pool of woody perennials and in below-ground biomass of grass are conservatively assumed to be constant and equal to the existing carbon stocks measured at the start of the project activity.

Since all pre-project living trees on lands to be planted in the proposed afforestation and reforestation CDM project activity have been measured, the baseline net removals by sinks are estimated on average tree basis rather than per hectare basis. The formulae used for the calculations are as follows:

Baseline carbon stocks:

$B_{(t)} = (B_{A(t)} + B_{B(t)})N$

where

- $B_{(t)}$ is the carbon stock in the living biomass within the project boundary at time t in the absence of the project activity (tonnes of carbon)
- $B_{A(t)}$ is the carbon stock in above-ground biomass per tree at time t in the absence of the project activity (tonnes of carbon per tree)
- $B_{B(t)}$ is the carbon stock in below-ground biomass per tree at time t in the absence of the project activity (tonnes of carbon per tree)
- N is the number of pre-project living trees

Carbon stocks in the above-ground biomass:

$B_{A(t)} = M_{(t)} \times 0.5$

where

- $B_{A(t)}$ is the carbon stock in above-ground biomass per tree at time t in the absence of the project activity (tonnes of carbon per tree)
- $M_{(t)}$ is the above-ground biomass per tree at time t that would have occurred in the absence of the project activity (tonnes of dry matter per tree)
- 0.5 is the carbon fraction of dry matter (tonnes of carbon per tonne of dry matter)

Carbon stocks in below ground biomass:

$B_{B(t)} = M_{(t)} \times R \times 0.5$

where

- $B_{B(t)}$ is the carbon stock in below-ground biomass per tree at time t in the absence of the project activity (tonnes of carbon per tree)
- $M_{(t)}$ is the above-ground biomass per tree at time t that would have occurred in the absence of the project activity (tonnes of dry matter per tree)
- R is the root–shoot ratio
- 0.5 is the carbon fraction of dry matter (tonnes of carbon per tonne of dry matter)

Above-ground biomass:

$M_{(t)} = V_{(t)} \times BEF_{(t)} \times D_{(t)}$

where

- $M_{(t)}$ is the above-ground biomass per tree at time t that would have occurred in the absence of the project activity (tonnes of dry matter per tree)
- $V_{(t)}$ is the stem volume (m^3 per tree)
- $D_{(t)}$ is the wood density (tonnes of dry matter per m^3)
- $BEF_{(t)}$ is the biomass expansion factor from stem biomass to above-ground biomass (dimensionless)

In 369.87 ha of degraded sand-dune-affected land proposed for the small-scale afforestation and reforestation CDM project activity, *Prosopis cineraria* is found in the largest number, which amounts to 435 trees out of the total 492 trees in the baseline. The above-ground biomass at time t, $M_{(t)}$, has been estimated using average biomass stock and growth rates specific to the region provided by the Haryana Forest Department. Since the age of the trees could not be estimated, an average mean annual increment (MAI) of 0.01752 m^3 per tree per year (this gives 8.5 m^3 per year for 492 trees) has been used to project the growth.[5] The MAI has been used to obtain the volume for each year. From the volume, the carbon stocks have been estimated using the formulae from the applied methodology.

[5] Letter no. 280, Indira Gandhi Canal Project, Forest Department, Government of Rajasthan

Table A10.12 Estimation of baseline net GHG removals by sinks (tonnes of CO_2 per year)

Year[a]	Trees[b]					
	AGB (tonnes of carbon)	BGB (tonnes of carbon)	Woody perennials and grasses[c] (tonnes of carbon)	Total carbon stock (tonnes of carbon)	Change in carbon stock (tonnes of carbon per year)	Baseline net GHG removal by sinks (tonnes of CO_2 equivalent per year)
A	*B*	*C*	*D*	$E = B + C + D$	$F = E_i - E_{i-1}$	$G = F \times 44/12$
2007	200.2	54.05	5.5	259.8	0	0
2008	209.4	56.54	5.5	271.5	12	43
2009	218.7	59.04	5.5	283.2	12	43
2010	227.9	61.53	5.5	295	12	43
2011	237.2	64.03	5.5	306.7	12	43
2012	246.4	66.53	5.5	318.5	12	43
2013	255.6	69.02	5.5	330.2	12	43
2014	265	71.52	5.5	341.9	12	43
2015	274	74.02	5.5	353.7	12	43
2016	283	76.51	5.5	365.4	12	43
2017	293	79.01	5.5	377.2	12	43
2018	302	81.5	5.5	388.9	12	43
2019	311	84	5.5	400.7	12	43
2020	320	86.5	5.5	412.4	12	43
2021	330	88.99	5.5	424.1	12	43
2022	339	91.49	5.5	435.9	12	43
2023	348	93.98	5.5	447.6	12	43
2024	357	96.48	5.5	459.4	12	43
2025	367	98.98	5.5	471.1	12	43
2026	376	101.47	5.5	482.8	12	43
2027	385	103.97	5.5	494.6	12	43

AGB – above-ground biomass; BGB – below-ground biomass

Notes:

[a] The starting year of the project is 2008. The year 2007 has been shown in the tables so as to illustrate the change in the carbon stock.

[b] Pre-project trees that existed before the starting of the project.

[c] Woody perennials and grasses constitute all the vegetation and shrubs, excluding the trees that exist in the area before the start of the project. Since the woody perennials are expected to decrease in the absence of the project, the carbon stocks in them are assumed to be constant and equal to the carbon stocks at the beginning of the project.

C.2 Estimate of the actual net GHG removals by sinks:

There are 492 pre-project trees allowed to remain on the project site. For the purpose of conservation, in the project scenario, the carbon stocks of the pre-project trees are assumed to be constant and the woody perennials and grass are assumed to be died out.

The formulae that have been given in the section III of the simplified baseline and monitoring methodologies for small-scale afforestation and reforestation project activities under the clean development mechanism implemented on grassland or cropland (AR-AMS0001; version 04.1) has been used for calculating the actual net GHG removals by sinks (ex ante).

The carbon stocks for the project scenario at the starting date of the project activity 4 (t = 0) is the same as the baseline stocks of carbon at the starting date of the project (t = 0).

For all other years, the carbon stocks of planted trees within the project boundary [$N(t)$] at time t has been be calculated as following:

$$N(t) = \sum_{i=1}^{l} (N_A(t)_i + N_{B(t)i})\, A_i$$

where

- $N_{(t)}$ is the total carbon stock in biomass at time t under the project scenario (tonnes of carbon)
- $N_{A(t)i}$ is the carbon stock in above-ground biomass at time t of stratum i under the project scenario (tonnes of carbon per hectare)
- $N_{B(t)i}$ is the carbon stocks in below-ground biomass at time t of stratum i under the project scenario (tonnes of carbon per hectare)
- A_i is the project activity area of stratum i (ha)

Above-ground biomass:

For above-ground biomass, $N_{A(t)}$ per stratum i is calculated as follows:

$$N_{A(t)i} = T_{(t)i} \times 0.5$$

where

- $N_{A(t)i}$ is the carbon stock in above-ground biomass at time t under the project scenario (tonnes of carbon per hectare)
- $T_{(t)i}$ is the above-ground biomass at time t under the project scenario (tonnes of dry matter per hectare)

- 0.5 is the carbon fraction of dry matter (tonnes of carbon per tonne of dry matter)

Volume and yield tables were used for the calculations. Hence, according to the methodology,

$$T_{(t)i} = S_{V(t)i} \times BEF \times WD$$

where

- $T_{(t)i}$ is the above-ground biomass at time t under the project scenario (tonnes of dry matter per hectare)
- $S_{V(t)i}$ is the stem volume at time t for the project scenario (m^3/ha)
- BEF is the biomass expansion factor (over bark) from stem to total above-ground biomass (dimensionless)
- WD is the basic wood density (tonnes of dry matter per m^3)

Below-ground biomass:

For below-ground biomass, $N_{B(t)}$ per stratum i is calculated as follows:

$$N_{B(t)i} = T_{(t)} \times R \times 0.5$$

where

- $N_{B(t)i}$ is the carbon stock in below-ground biomass at time t under the project scenario (tonnes of carbon per hectare)
- $T_{(t)}$ is the above-ground biomass at time t under the project scenario (tonnes of dry matter per hectare)
- R is the root–shoot ratio (tonnes of dry matter/tonnes of dry matter)
- 0.5 is the carbon fraction of dry matter (tonnes of carbon per tonne of dry matter)

The plantation has been planned in block mixtures, each block having a particular species. The GHG removal by each of the seven species has been calculated. The volume has been calculated from the local volume tables[6] provided by the Haryana Forest Department. In all the cases, to make conservative estimation, only the rate of growth of trees of area having physiographic and edaphic factors similar to the project site has been taken. The MAIs of the tree species are 0.0333 m^3 for Eucalyptus hybrid, 0.00156 m^3 for *Ailanthus excelsa,* 0.0022 m^3 for *Acacia tortilis,* 0.0144 m^3 for *Dalbergia sissoo,* 0.01752 m^3 for *Prosopis cineraria,* and 0.0224 m^3 for *Acacia nilotica* and *Zizyphus mauritiana.*

[6] Reference: letter no. 280, Indira Gandhi Canal Project, Forest Department, Government of Rajasthan

The BEF of 3.4 has been taken from Table 3A.1.10 of the *IPCC Good Practice Guidance for Land use, Land-use Change, and Forestry.* It has been chosen from the default values in the absence of any national data in this regard. The default value is that of a tropical broadleaf forest having a BEF in the range of 2.0 to 9.0. The wood density has been obtained from the table of *IPCC Good Practice Guidance for Land use, Land-use Change, and Forestry* (refer to Table A10.9). The root–shoot ratio of 0.27 (Table 3A.1.8 of *IPCC Good Practice Guidance for Land use, Land-use Change, and Forestry*) has been used to calculate the BGB because no national data are available.

The plantation of trees shall be divided into short rotation (of 10 years) and medium rotation (of 20 years) and fruit trees. The biomass has been calculated from the volumes obtained from the tables and using the value of the wood density. Replanting shall be taken up in the first three years. Still mortality of 20% has been taken to make the estimations conservative.

Project Emissions

In 2008, plantation shall be undertaken in the 369.87 ha selected for the project with 1000 trees in every hectare. The saplings shall be transported in bullock carts to the respective planting area from the nurseries that have been made at the villages itself. Trees shall be planted in 45 cm × 45 cm × 45 cm pits in which 1 kg organic manure shall be spread, amounting to an application of 1 tonne of fertilizer in every hectare. This manure shall be collected from locally available sources and transported using bullock carts, causing zero emission in the transportation of the manure. Manure shall be applied only in the year of plantation. The GHG emissions from the fertilizers are 0.92% of the annual GHG sequestered by the project. The net nitrogen content of organic manure is 5% for dried matter (IPCC Emission Factor Data Base). N_2O emissions are 1% of the total nitrogen content (Executive Board report 33, AR methodology tool). Since the emissions are below 10% of the net GHG sequestered, they have been taken as zero. The ex-ante actual GHG removal by sinks is calculated as follows:

$$\Delta C_{ACTUAL} = \Delta C_{PROJ,t} - GHG_{PROJ,t}$$

where

- ΔC_{ACTUAL} is the ex-ante actual GHG removal by sinks (tonnes of CO_2 equivalent per year)

- $\Delta C_{PROJ,t}$ is the project GHG removals by sinks (tonnes of CO_2 per year)
- $GHG_{PROJ,t}$ is the project emissions (tonnes of CO_2 equivalent per year)

Table A10.13 Carbon stock in above-ground biomass (tonnes of carbon)

Year	Pre-project vegetation (tonnes of carbon)	Project strata	Total carbon stock in above-ground biomass (tonnes of carbon)	S1 (tonnes of carbon	S2 (tonnes of carbon)	S3 (tonnes of carbon)	S4 (tonnes of carbon)	S5 (tonnes of carbon)	S6 (tonnes of carbon)
2007	205.7	0	0	0	0	0	0	0	205.7
2008	200.2	21.6	29.0	69.0	44.0	49.5	136.1	43.1	592.5
2009	200.2	43.1	58.1	138.0	87.9	99.0	272.1	86.3	984.7
2010	200.2	64.7	87.1	207.0	131.9	148.5	408.2	129.4	1,377.0
2011	200.2	86.2	116.2	276.0	175.8	198.0	544.2	172.6	1,769.2
2012	200.2	107.8	145.2	345.0	219.8	247.5	680.3	215.7	2,161.5
2013	200.2	983.4	208.7	446.2	462.1	574.9	928.1	388.3	4,191.9
2014	200.2	1,859.0	272.1	547.4	704.4	902.2	1,175.8	560.9	6,222.0
2015	200.2	2,734.6	335.5	648.7	946.7	1,229.6	1,423.6	733.4	8,252.3
2016	200.2	3,610.2	399	749.9	1,189.0	1,556.9	1,671.4	906.0	10,282.6
2017	200.2	4,485.7	462.4	851.1	1,431.3	1,884.2	1,919.2	1,078.6	12,312.7
2018	200.2	21.6	29	69.0	1,729.1	2,787.2	2,593.4	1,423.7	8,853.2
2019	200.2	43.1	58.1	138.0	2,027.0	3,690.1	3,267.6	1,768.9	11,193.0
2020	200.2	64.7	87.1	207.0	2,324.8	4,593.1	3,941.7	2,114.0	13,532.6
2021	200.2	86.2	116.2	276.0	2,622.6	5,496.0	4,615.9	2,459.2	15,872.3
2022	200.2	107.8	145.2	345.0	2,920.4	6,399.0	5,290.1	2,804.3	18,212.0
2023	200.2	983.4	208.7	446.2	3,913.8	8,422.9	6,895.4	3,537.8	24,608.4
2024	200.2	1,859.0	272.1	547.4	4,907.2	10,446.8	8,500.7	4,271.2	31,004.6
2025	200.2	2,734.6	335.5	648.7	5,900.6	12,470.7	10,105.9	5,004.7	37,400.9
2026	200.2	3,610.2	399	749.9	6,894.0	14,494.5	11,711.2	5,738.1	43,797.1
2027	200.2	4,485.7	462.4	851.1	7,887.5	16,518.4	13,316.5	6,471.6	50,193.4

Notes:

1. See Table A10.8 for the species of each stratum
2. Pre-project vegetation includes the above-ground biomass of trees and woody perennials (shrubs). The carbon stocks of woody perennials are included in the above-ground biomass alone. The woody perennials are expected to be removed during the site preparation but the carbon stocks of pre-project vegetation are otherwise taken to be constant for a conservative estimation.

Table A10.14 Total below-ground biomass carbon stock (tonnes of carbon dry matter) Years

	Pre-project vegetation (tonnes of carbon)	Project strata							Total carbon stock in below-ground biomass (tonnes of carbon)
		S1 (tonnes of carbon)	S2 (tonnes of carbon)	S3 (tonnes of carbon)	S4 (tonnes of carbon)	S5 (tonnes of carbon)	S6 (tonnes of carbon)	S7 (tonnes of carbon)	
2007	54.1	0	0	0	0	0	0	0	54.1
2008	54.1	5.8	7.8	18.6	11.9	13.4	36.7	11.6	159.9
2009	54.1	11.6	15.7	37.3	23.7	26.7	73.5	23.3	265.9
2010	54.1	17.5	23.5	55.9	35.6	40.1	110.2	34.9	371.8
2011	54.1	23.3	31.4	74.5	47.5	53.5	146.9	46.6	477.8
2012	54.1	29.1	39.2	93.1	59.3	66.8	183.7	58.2	583.5
2013	54.1	265.5	56.3	120.5	124.8	155.2	250.6	104.8	1,131.8
2014	54.1	501.9	73.5	147.8	190.2	243.6	317.5	151.4	1,680.0
2015	54.1	738.3	90.6	175.1	255.6	332.0	384.4	198.0	2,228.1
2016	54.1	974.7	107.7	202.5	321.0	420.4	451.3	244.6	2,776.3
2017	54.1	1,211.2	124.8	229.8	386.5	508.7	518.2	291.2	3,324.5
2018	54.1	5.8	7.8	18.6	466.9	752.5	700.2	384.4	2,390.3
2019	54.1	11.6	15.7	37.3	547.3	996.3	882.2	477.6	3,022.1
2020	54.1	17.5	23.5	55.9	627.7	1,240.1	1,064.3	570.8	3,653.9
2021	54.1	23.3	31.4	74.5	708.1	1,483.9	1,246.3	664.0	4,285.6
2022	54.1	29.1	39.2	93.1	788.5	1,727.7	1,428.3	757.2	4,917.2
2023	54.1	265.5	56.3	120.5	1,056.7	2,274.2	1,861.8	955.2	6,644.3
2024	54.1	501.9	73.5	147.8	1,324.9	2,820.6	2,295.2	1,153.2	8,371.2
2025	54.1	738.3	90.6	175.1	1,593.2	3,367.1	2,728.6	1,351.3	10,098.3
2026	54.1	974.7	107.7	202.5	1,861.4	3,913.5	3,162.0	1,549.3	11,825.2
2027	54.1	1,211.2	124.8	229.8	2,129.6	4,460.0	3,595.5	1,747.3	13,552.3

Notes:

1. See Table A10.8 for the species of each stratum
2. Pre-project vegetation consists of the below-ground carbon stocks of the 492 pre-project trees alone and not woody perennials as their carbon stocks has been assumed to be above ground only. The stocks are assumed to be constant for conservative estimations.

Table A10.15 Actual net GHG removals by sinks

Year	AGB[a] (tonnes of carbon)	BGB[b] (tonnes of carbon)	Total carbon stocks (tonnes of carbon)	Change in carbon stocks (tonnes of carbon per year)	Actual net GHG removal by sinks (tonnes of CO_2 equivalent per year)
A	*B*	*C*	*D = B + C*	*E = D2 – D1*	*F = E × 44/12*
2007	205.7	54.1	259.8	0	0
2008	592.5	159.9	752.4	492.6	1,806.0
2009	984.7	265.9	1,250.5	498.2	1,827.0
2010	1,377.0	371.8	1,748.7	498.2	1,827.0
2011	1,769.2	477.8	2,246.9	498.2	1,827.0
2012	2,161.5	583.5	2,745.0	498.2	1,827.0
2013	4,191.9	1,131.8	5,323.5	2,578.4	9,454.0
2014	6,222.0	1,680.0	7,901.9	2,578.4	9,454.0
2015	8,252.3	2,228.1	10,480.4	2,578.4	9,454.0
2016	10,282.6	2,776.3	13,058.8	2,578.4	9,454.0
2017	12,312.7	3,324.5	15,637.2	2,578.4	9,454.0
2018	8,858.2	2,390.3	11,243.6	–4,393.7	–16,110.0
2019	11,193.0	3,022.1	14,215.0	2,971.4	10,895.0
2020	13,532.6	3,653.9	17,186.4	2,971.4	10,895.0
2021	15,872.3	4,285.6	20,157.8	2,971.4	10,895.0
2022	18,212.0	4,917.2	23,129.2	2,971.4	10,895.0
2023	24,608.4	6,644.3	31,252.5	8,123.3	29,785.0
2024	31,004.6	8,371.2	39,375.8	8,123.3	29,785.0
2025	37,400.9	10,098.3	47,499.0	8,123.3	29,785.0
2026	43,797.1	11,825.2	55,622.3	8,123.3	29,785.0
2027	50,193.4	13,552.3	63,745.6	8,123.3	29,785.0

AGB – above-ground biomass; BGB – below-ground biomass

Notes:

[a] AGB total as demonstrated in Table A10.13

[b] BGB total as demonstrated in Table A10.14

C.3 Estimated leakage:

The project area is degraded/degrading cropland, which is generally left fallow, and profitable agriculture is carried out on an average interval of three years and only one crop is taken instead of two, which is the general trend in the nearby irrigated agricultural fields. Thus, the agriculture activity shifting to nearby area is insignificant and far below 10%.

The cattle in the area are stall fed. The cattle feed comprises agriculture waste as well as fodder crops grown in the adjoining fertile fields, which are canal irrigated. The region is visited by migratory cattle herd from Rajasthan from March to May. These herds feed on agricultural wastes and grasses growing on roadside and are not allowed to graze on the privately owned lands of the proposed project. Therefore, the number of displaced grazing animals due to the project activity is less than 10% of the average grazing capacity of the project area.

Therefore, according to paragraph 30 of AR-AMS0001/Version 4.1, there is no displacement of activities outside the project. So the estimated leakage is zero.

C.4 The sum of C.2 minus C.1 minus C.3 representing the net anthropogenic GHG removals by sinks of the proposed small-scale afforestation and reforestation CDM project activity:

The net anthropogenic GHG removals by sinks of the proposed small-scale afforestation and reforestation CDM project activity have been estimated using formula 21, paragraph 33 of AR-AMS0001/ Version 4.1 as follows:

$$ER_{\mathrm{AR\ CDM},t} = \mathrm{C}_{\mathrm{PROJ},t} - \mathrm{C}_{\mathrm{BSL},t} - GHG_{\mathrm{PROJ},t} - L_t$$

where

- $ER_{\mathrm{AR\ CDM},t}$ is the net anthropogenic GHG removals by sinks (tonnes of CO_2 equivalent per year)
- $\mathrm{C}_{\mathrm{PROJ},t}$ is the project net GHG removals by sinks (tonnes of CO_2 equivalent per year)
- $\mathrm{C}_{\mathrm{BSL},t}$ is the baseline net GHG removals by sinks (tonnes of CO_2 equivalent per year)
- $GHG_{\mathrm{PROJ},t}$ is the project emissions (tonnes of CO_2 equivalent per year)
- L_t is the leakage (tonnes of CO_2 equivalent per year)

The project emissions have been deducted from the project GHG removals by sinks as shown in Table A10.16. The net anthropogenic GHG removals by sinks of the proposed small-scale afforestation and reforestation CDM project activity have been estimated to be 231,920 tonnes of CO_2 equivalent.

C.5 Table providing values obtained when applying equations from the approved methodology:

Table A10.16 Net anthropogenic GHG removals by the project years

Year	Estimation of baseline net GHG removals by sinks[a] (tonnes of CO_2 equivalent)	Estimation of actual net GHG removals by sinks[b] (tonnes of CO_2 equivalent)	Estimation of leakage[c] (tonnes of CO_2 equivalent)	Estimation of net anthropogenic GHG removals by sinks (tonnes of CO_2 equivalent)
2008	43	1,806	0	1,763
2009	43	1,827	0	1,784
2010	43	1,827	0	1,784
2011	43	1,827	0	1,784
2012	43	1,827	0	1,784
2013	43	9,454	0	9,411
2014	43	9,454	0	9,411
2015	43	9,454	0	9,411
2016	43	9,454	0	9,411
2017	43	9,454	0	9,411
2018	43	–16,110	0	–16,153
2019	43	10,895	0	10,852
2020	43	10,895		10,852
2021	43	10,895	0	10,852
2022	43	10,895	0	10,852
2023	43	29,785	0	29,742
2024	43	29,785	0	29,742
2025	43	29,785	0	29,742
2026	43	29,785	0	29,742
2027	43	29,785	0	29,742
Total				231,920

GHG – greenhouse gas

Notes:

[a] Refer Table A10.12

[b] Refer Table A10.15

[c] Refer Section C.3

SECTION D: ENVIRONMENTAL IMPACTS OF THE PROPOSED SMALL-SCALE AFFORESTATION AND REFORESTATION CDM PROJECT ACTIVITY

D.1 Provide analysis of the environmental impacts, including trans-boundary impacts (if any):

The proposed small-scale afforestation and reforestation CDM project activity is expected to have the following environmental impacts through its afforestation activities:

- **Sand dune fixation and controlling wind erosion:** The major soil type in the project area is desert soil with a high sand content, and the area is affected by shifting sand dunes and frequent wind erosion/dust storms. Due to the lack of vegetation cover, mild to heavy winds often occurring in this region blow up clouds of dust and sediments into the atmosphere and denudes the adjacent area of vegetation. As a result, these lands have severely degraded and are degrading. Therefore, establishment of plantations in 369.87 ha of the project area under the proposed small-scale afforestation and reforestation CDM project activity will help stabilize the sand dunes in the project area and arrest soil erosion. An environment impact study in Haryana Community Forestry Project (HCFP) villages indicated that with the establishment of plantation by HCFP, moving sand dunes have been fixed and dust storms have been mitigated in 25%–70% of the HCFP villages.
- **Improving land fertility:** The productivity of these lands is very low and highly depends on rainy event due to inaccessible irrigation system and low soil fertility. The proposed afforestation activity will enhance the soil moisture and humus content by improving nutrient cycling and soil water regime. Soil sampling and analyses in 11 HCFP villages indicated that the soil organic carbon (0–30 cm) of the five- to six-year-old plantation established by the project ranges from 0.43% to 1.29% compared to 0.31%–1.21% in the open lands adjacent to the plantation.
- **Sheltering the surrounding productive croplands**: Shifting sand dunes threaten croplands adjacent to the project lands and make them less productive. The loss of cropland due to shifting

sand dunes has been witnessed in many villages. Restoration of these degraded areas will lead to positive impacts in the areas outside the project boundary, especially the fertile agricultural areas, since the afforestation activities in the project activity area will contribute to stabilize shifting sand dune and alleviate wind erosion and thus provide shelter for the neighbouring fertile fields. An environment impact study in HCFP villages indicated that with the establishment of plantations and planting of moving sand dunes, the loss of croplands due to shifting sand dunes has been brought under control in some HCFP villages.

- **Risk analysis and countermeasures**:
  - **Site preparation**: Site preparation has the potential to disturb the vegetation and soil in the planting sites. The main technical measures to be employed to mitigate the impacts designed in the project are to plant the trees with low density (1000 trees per hectare), limited pit size (45 cm × 45 cm × 45 cm), and retaining all the existing vegetation. Even the castor crop that is proposed to be taken in the first two years shall be raised only by dibbling seeds and not more than 10 cm^2 of soil at each dibble point will be disturbed. As a result, the surface area disturbed by site preparation is estimated to account for less than 3% of the total land surface. Therefore, the site and soil preparation will have minor negative impacts on original soil and vegetation.
  - **Fertilization**: In the proposed small-scale afforestation and reforestation CDM project activity, only organic manure will be applied within the small planting pits rather than through dispersal over the entire land, thus leading to maximum impact on the plant while causing least disturbance outside.
  - **Pesticide**: Under the proposed small-scale afforestation and reforestation CDM activity, no pesticides are proposed to be used as a preventive measure. Only hygienic measures to control pests and diseases will be adopted. Only in the case of a severe outbreak of pests, the use of pesticides would be considered and then suitable safeguards against

the environmental effects of the pesticides would be undertaken to ensure that residues do not escape into water sources for the people and cattle.

- **Fire risk**: Fires in the plantations escaping into the neighbourhood is a potential risk of the proposed small-scale afforestation and reforestation CDM project activity. However, the Haryana Forest Department shall reduce this risk through awareness and training to local farmers/ communities in collaboration with the society.
- **Invasives**: Native tree species are being planted in the project area apart from Eucalyptus hybrid, which has been raised in the district by the Haryana Forest Department under various forestry projects as well as by farmers in their private lands for the past many decades and there is no evidence of this species becoming invasive. Given the countermeasures to be implemented, all the aforementioned risks are not considered significant by the project participants.

D.2 If any negative impact is considered significant by the project participants or the host party, a statement that project participants have undertaken an environmental impact assessment, in accordance with the procedures required by the host party, including conclusions and all references to support documentation:

Environmental impact assessment is not required for the proposed small-scale afforestation and reforestation CDM project activity, since this project does not have any significant negative impacts on the environment.

D.3 Description of planned monitoring and remedial measures to address significant impacts referred to in D.2 earlier:

Not applicable

SECTION E: SOCIO-ECONOMIC IMPACTS OF THE PROPOSED SMALL-SCALE AFFORESTATION AND REFORESTATION CDM PROJECT ACTIVITY

E.1 Provide analysis of the socio-economic impacts, including trans-boundary impacts (if any):

The project participants belong to communities whose main income source is farming and animal husbandry. These farmers have small land holdings, a considerable part of which is degraded and degrading croplands and not suitable for profitable agriculture as investment costs are too high for them to carry out any kind of farming or planting activities. Only 35% of the lands owned by the farmers are productive and give them an annual income of ₹4000 per acre, while their earnings from animal husbandry range around ₹1000–5000 a year. Data[7] on the land details and cattle holding of the farmers participating in the proposed small-scale afforestation and reforestation CDM project activity were collected to estimate their average annual income individually as well as for all the villages and households (as shown in Table A10.17). The data show that 92% of the households are below the international poverty norm of $1.00 per day and have an average daily income of ₹18.

Field visits and interviews with the project participants have shown that the households of majority of the farmers lack basic amenities such as running water and bathroom facilities. Most of the farmers reside in small mud houses and the cattle are also housed in the same dwelling. Fuelwood is the sole medium of cooking in all the households, which puts a lot of burden on the women as they have to walk long distances for fuelwood collection. They do not own vehicles for transport, and they either walk long distances or use animal transport. It was learnt through repeated interactions with the project participants and their family members that they do not have any savings or assets apart from the land and a few cattle that they possess. The average annual income of these households varies from ₹20,000 to ₹40,000, which is still below the international poverty ceiling of around ₹86,400. A small percentage of farmers (8%) having larger land holdings have also been included in the project to make the activity more viable and to enable smooth and efficient functioning of the proposed small-scale afforestation and reforestation CDM project activity.

---

[7] Data sheets containing land and cattle holding details along with the individual annual income of the project participants will be made available to the DOE.

Table A10.17 Mean annual income of the participants

Project village	Total number of household participants	Average annual income per capita (₹)
Neemla	43	5,594
Dhani Sheranwali	30	7,074
Poharkan	19	11,552
Mallekan	17	5,081
Madho Singhana	20	8,768
Gudia Khera	43	5,265
Bhuratwala	17	8,564
Umedpura	38	5,119
Total	227	6,610

Notes:

1. Income pattern: 35% and 65% of land are productive and unproductive, respectively.
2. Income from productive area is ₹4000/acre/annum.
3. Income from unproductive area is ₹1500/acre/annum.
4. Income from buffaloes is ₹2000/annum/buffalo.
5. Income from cows is ₹1000/annum/cow.
6. Number of members per household is 6.1.

Ninety two per cent of the project participants have an annual income that places them far below the international poverty ceiling and their households lack basic running water and toilet facilities. Their present socio-economic conditions clearly place them as low-income communities. Therefore, to maximize the socio-economic benefits from the proposed small-scale afforestation and reforestation CDM project activity, a series of PRA methods were adopted in consulting with the farmers to understand their expectations and preferences, wishes and concerns, so that the proposed project activity would better respond to their desire for livelihood development (see Section F). It is expected that 227 farmers from eight villages of three developmental blocks will benefit from the proposed small-scale afforestation and reforestation CDM project activity. The main socio-economic benefits of the project include the following:

- **Income generation:** The project will open up certain new economic opportunities and enhance the income of farmers through timber and fruit production besides earnings from the sale of carbon credits. About 227 local farmers of eight villages will benefit from the project. The timber produced is expected to fetch a lucrative price in the local markets. Besides, ber (*Ziziphus mauritiana*) is the most preferred fruit species in the project area. The farmers can grow this fruit tree and realize a good market price at the farm gate as well as in the local markets. Since it is the first CDM forestry project in the country, it is expected to provide excellent opportunities for training and exposure to this fast-developing field and at least some educated young men and women in these villages hope to be able to access entirely new employment avenues in the field of CDM.

- **Sustainable fuelwood supply:** The local communities depend on fuelwood for their living to a certain extent, and the proposed small-scale afforestation and reforestation CDM activity will provide more sustainable fuelwood resources to the local farmers. This will make the living conditions of some womenfolk who have to walk long distances for fuelwood collection easier.
- **Strengthening social cohesion**: Individual farmer households/ communities are unable to successfully manage the investment, production, and market chain of the timber and non-wood forest products, which normally takes a much longer period than food production. In addition, the lack of organizational instruments also prevents them from overcoming technological barriers. Overall, the proposed small-scale afforestation and reforestation CDM project activity will entail close interaction between individuals, communities, farm forestry, and local government, with intensified communication among them, thus supporting the network for social and productive services.
- **Social well-being:** The proposed small-scale afforestation and reforestation CDM project activity will generate additional employment through site preparation, plantation, and other intercultural operations. Besides, the project will enhance the community bondage and elevate the status of women by ensuring their participation in the proposed project activity. Since this is the first CDM forestry project in India, and only second in the world, it is expected to bring recognition to the community working for the project and thus contribute significantly to the sense of well-being.
- **Better agricultural production:** It is expected that the afforestation activities of the project will lead to fixation of shifting sand dunes.

The proposed small-scale afforestation and reforestation CDM project activity will also increase the water-holding capacity of the soil and add humus to the soil, which would in turn make agriculture more viable in the neighbourhood.

- **Technical training and demonstration:** Interview with local communities indicated that local farmers/communities are usually short of access to quality seed sources and lack skills for producing high quality seedlings for successful tree planting, as well as for preventing planted trees from impact of fire, pest, and diseases. This is one of the important barriers of local communities in planting trees on their lands. In the proposed small-scale afforestation and reforestation CDM project activity, the local forestry agencies as well as large farms will organize the training for local communities to assist them in understanding and evaluating the issues of hosting the proposed project activity, both on-site and off-site, such as seed and seedling selection, nursery management, site preparation, and planting models.

The proposed small-scale afforestation and reforestation CDM project activity is, therefore, expected to uplift the socio-economic status of the project participants and the communities at large by meeting the various objectives mentioned earlier. The potential socio-economic risks and countermeasures are as follows:

- **Cultural resources:** There are no cultural relics and/or cultural reserves that have been identified in the project area, and consequently, no damage to non-replicable cultural property will occur under the proposed small-scale afforestation and reforestation CDM project activity. Meanwhile, the project does not involve any site used for local social gatherings or other spiritual activities. Thus, the proposed project activity will not impact the normal local gatherings and religious activities.
- **Economic risk:** The potential economic risks will be poor management of the plantations established under the project such as lack of pest and fire control measures, which could contribute to project failure and farmers' loss. The proposed small-scale afforestation and reforestation CDM project activity has been designed to mitigate this risk by providing technical assistance and training to farmers and communities, by farm forestry and the extension network of the forestry sector. The local government

forestry office, which is experienced in reforestation and forest management, will provide the technical assistance to the farmers/ communities.

- **Possibility of low value of carbon credits:** This is almost the lowest risk factor, as there is increasing realization of the threat of global warming and the countries that were reluctant to join the Kyoto Protocol till recently are coming on board. It is expected that the demand for carbon credits will keep on increasing, thereby helping raise their prices in the future. None of these risks are considered significant.

E.2 If any negative impact is considered significant by the project participants or the host party, a statement that project participants have undertaken a socio-economic impact assessment, in accordance with the procedures required by the host party, including conclusions and all references to support documentation:

The project participants or the host party did not find any significant negative impact.

E.3 Description of planned monitoring and remedial measures to address significant impacts referred to in E.2:

Not available

SECTION F: STAKEHOLDERS' COMMENTS

F.1 Brief description of how comments by local stakeholders have been invited and compiled:

To collect the relevant information for the "Small Scale Cooperative Afforestation CDM Pilot Project Activity on Private Lands Affected by Shifting Sand Dunes in Sirsa, Haryana", a series of PRA exercises were conducted. The main objectives of this PRA exercise were:

- To know the aspirations of the participants/villagers on small-scale afforestation and reforestation CDM project activities and identify project participants.
- To evaluate the eligibility of the lands under the project area.
- To analyse the potentiality of additionality and leakage in the project area.

- To identify the discrete parcels to be included under the project area.
- To explore the crop productivity, grazing pattern, climate, and other relevant points.
- To know the income sources of the project participants and study the socio-economic conditions of the households and the communities in the eight villages.
- To collect field data on individual land details and cattle holdings of the project participants.

Tools Used

A series of open meetings were conducted by using the following tools:

- Semi-structured interviewing
- Focus group discussions
- Village and resource mapping
- Seasonal/crop cycle mapping
- Timeline analysis
- Transect walk

A PRA exercise is in progress in one of the project villages. Detailed PRA exercises were conducted for eight participating villages: Neemla, Dhani Sheranwali, Bhuratwala, Poharkan, Umedpura, Mallekan, Madho Singhana, and Gudia Khera. The Haryana Forest Department informed the villagers well in advance about the visit of the PRA team. The team met the farmers in their respective villages at a common place. At each village, activities such as preparation of village and resource map, timeline analysis, and crop cycle/seasonal maps were performed. The villagers were clearly informed about the society by-laws and the criteria and eligibility for the CDM project. They were particularly asked questions under the following broad categories, which were captured in video and camera as evidence/proof for records to be sent along with PDD during project submission.

Land eligibility: For ascertaining land eligibility criteria of the CDM project activities, villagers were asked about their proposed land status, whether it was a forest area, and if so, since when.

Grazing for leakage criteria: Villagers were specifically asked about grazing of their cattle in the proposed land and also any grazing by migratory cattle from adjoining areas.

Agriculture: Villagers were asked about the agriculture of crop in their proposed land, which are degraded sand dune areas.

Timeline and history tracing: Villagers were asked to elaborate about their areas and record any developmental activities in their areas since 1940.

Income: Land and cattle holding data collected by the field staff was cross-checked with the individual participants, and the participants were interviewed about the productivity and yield of their lands.

Alternative to land: Villagers were asked to tell whether they have any other alternative available for the project lands, had not this project been suggested.

Species choice: Villagers were categorically asked to tell about the choice of tree species selected for planting with their full consent and choice.

Seasonality of crops: Villagers were requested to prepare a crop cycle map showing seasonality of various crops in their area.

A total of 227 farmers from all the eight selected villages finally volunteered to give their degraded land, which constitutes 369.87 ha (as per GPS calculations), for the project activity.

F.2 Summary of the comments received:

Farmers' Understanding on CDM and its Benefits

Initially, a thorough discussion on CDM was organized with the farmers. It was stated that afforestation and reforestation CDM project activities could be carried out if a certain amount of the land is degraded in nature or not suitable for growing profitable agricultural crops. The team stated that there could be several benefits, which are listed as follows, from the proposed small-scale afforestation and reforestation CDM project activity:

- Growing timber or fruit crops in the degraded lands, which were otherwise kept fallow and not utilized for agricultural practices.
- Providing timber after 10–20 years interval.

- Providing annual fruits after five years onwards.
- Earning carbon credit, besides conservation of land and environment, and development of local area.

During the PRA exercise, the farmers were made aware of the eligibility conditions of lands under the small-scale afforestation and reforestation CDM methodology. The various clauses of land eligibility were discussed with the farmers. They were informed that the lands to be afforested should preferably be degraded or marginal agricultural lands. Most of the questions regarding the suitability of lands were centred on why such lands are to be selected for the project.

The farmers were informed that for getting all the aforesaid benefits, they might contribute their degraded lands at least for 20 years (minimum) or for 30 years (maximum). At the end of the interaction, farmers (80% of the total respondents) expressed their willingness to contribute their lands for 20 years, as they found a 30-year time period too long.

During the interactions with the villagers, it was found that farmers were willing to take up this project. The active participation of the farmers in the discussion and their keen interest in knowing detailed specifics of the project are proof for this. All the farmers who participated in the activity volunteered to contribute lands for the project and were quite happy in being allowed to manage the plantation by them. However, the farmers insisted that each village should be represented in the plantation management committee. This was noted, and it was decided that representation from each village shall be ensured.

The draft by-laws were discussed with the farmers in the brain-storming exercise, and it was felt that farmers were willing to take part in the afforestation activity on their private lands. The PRA team elaborately explained to the villagers that a society would be established for this afforestation activity, which would be having two types of memberships, namely, general membership and executive membership. The membership of the society will be granted to any adult person of Ellenabad block and its neighbourhood who is willing to take part in the project by making eligible land under his/her lawful ownership available for meeting the objectives of the society. All such members shall constitute the general body of the society.

The society will accept only those lands that the owners are willing to commit to it for the proposed small-scale afforestation and reforestation CDM project activity for a minimum period of 20 years. The land should be at least 0.05 ha in size and should not be considered suitable for raising agricultural crops regularly on account of either low productivity or difficulty in management. During the PRA, the landowners agreed that they shall be responsible for the protection and management of the plantations raised on their lands subject to the overall supervision and satisfaction of the society. These lands should not be withdrawn during the project period of 20 years.

The villagers were informed that the general body of the society shall approve the programme of the activities of the society for the ensuing year and decide on the manner of sharing of benefits, including income from carbon credits. Besides, the farmers agreed that there will be an executive body consisting of minimum nine members with one each being elected from the eight participating villages and one member from the forest department as the secretary-cum-treasurer. At least one-third of the members of the executive body shall be women. In the event of inadequate representation of women, the patron (Haryana Forest Department) shall nominate such number of women from among the members of the society, or their families, as are necessary to fulfil this mandatory requirement.

The executive body shall be responsible for the day-to-day management of the society under the overall control and guidance of the general body. The executive body shall elect its president and vice president from among its elected and nominated members. The patron of the society shall appoint a serving forest officer not below the rank of the forest range officer as the secretary-cum-treasurer to the executive body. The society would have full right and control to decide on the proposed small-scale afforestation and reforestation CDM project activity and benefit sharing.

The farmers also agreed that once the land would be contributed for the proposed small-scale afforestation and reforestation CDM project activity, all the lands that have been considered for the project activity for 20 years would be managed by the landowner themselves under the overall supervision of the society, which would decide further activities on this project area. General members would continue to be the legal owner of their contributed land and possess the power to transfer or sell the land to a person who will be willing to accept all the by-laws of the society and thereby become its member.

Resource Inventory

For assessing the economic and social resources of the villages, PRA tools such as "village mapping and timeline analysis" were used. Farmers identified the following resources during the inventory analysis:

- **Land and water**: Village mapping exhibited that all the eight villages are located along the southernmost part of the Sirsa district. The fertile lands of these villages are under agricultural practices, which are irrigated by canal-fed irrigation system.

  There is no perennial river in the area, and Ghaggar is the only river flowing from north to west through the central part of the Sirsa district. During the rainy season, its water is diverted into the southern canal flowing through these villages. Apart from this, other sources of water are available only during the rains. There is a network of irrigation canals spread over these villages, originating from the Bhakra Canal, but are not present in the project area.

  Timeline analysis reveals that the areas of the proposed small-scale afforestation and reforestation CDM project activity were deforested during 1947–50, that is, the post-Partition (between India–Pakistan) period partially due to population growth and high dependency of local people on forest products for their livelihood and construction material for their houses, which lead to deforestation. The villagers identified the lands under the project as unproductive for agricultural crops as they are sandy and heavily affected by sand dunes. The area is affected by aeolian (wind-blown) sand.

  The villagers confirmed in PRA that the current land use of the project area is of degraded croplands. There are only a few shrubs or grasses in the area, which was cross-checked during the PRA.

  During the transect walk, farmers revealed (and the team also found) that the area has sand dunes (locally called tibbas), which cover the whole tract passing through the project area. These sand dunes (tibbas) found around Ellenabad are as high as 9 m (approximately) and are broad-based transverse ridges, a few of them even 3 km long.

- **Vegetation cover**: During the transect walk, it was found that the natural tree species available in the area are jand (*Prosopis cineraria*), beri (*Ziziphus mauritiana*), jaal (*Salvadora oleoides*), reru (*Acacia lucophloea*), kair (*Capparis decidua*), kikar (*Acacia nilotica*), and pipal (*Ficus religiosa*). It was reported that trees such as *Acacia nilotica, Ailanthus excelsa, Acacia tortilis, Albizzia lebbek*, and *Azadirachta indica* are vulnerable to frost. The old natural forests in these areas have vanished due to the increasing population over the years. Only a few trees and shrubs exist in this region, and the tree crown cover is not more than 5% (ocular estimate).
- **Agriculture**: It was evident from the discussions with the farmers during the PRA that there are large tracts of lands on which very limited agricultural practices are possible. Droughts prevail during October and November and in the summer months. As this region receives erratic and scanty rainfall (25–50 cm), the farmers get some yield from these lands on an average of once in three–four years (the farmers discussed with the PRA team that in their area, good rainfall is a rarity as scanty rainfall occurs only once in three–four years). There are a few irrigation canals near the selected villages, but these canal systems are not available in the project area. As rainfall in these lands is scanty, even the rain-fed crop productivity is very low. The lands having access to canal-fed irrigation supports normal agriculture. About 30%–40% of the lands owned by the farmers are productive. The main cropping pattern is two-crop systems. Paddy, cotton, and bajra crops are grown in the *Kharif* season (*Kharif* means autumn harvest in India, which is also known as summer or monsoon crops), whereas wheat, gram, barley, rapeseed, and mustard are grown in the *Rabi* season (*Rabi* means spring harvest in India, which is also known as winter crop). Cotton, oilseeds, and citrus (kinnow) are the main cash crops of this area practised in canal-fed irrigated areas. These areas give the farmers an annual income

of ₹4000 per acre, while the income from the degraded areas ranges anywhere around ₹1000–1500 per acre annually.

The farmers were also asked whether they presently have or are considering any options related to any other use of these lands. It was learnt from the farmers that they do not have any other alternative use of these lands and feel that by afforesting these degraded lands under the proposed small-scale afforestation and reforestation CDM project activity, they would obtain means of additional income from the wood and non-wood products and will also earn money from the sale of carbon credits (through the sale of tCERs). The farmers expect that the timber produced through the proposed CDM project activity would also fetch a good market price.

- **Livestock and grazing**: During the PRA exercise, information regarding livestock and grazing pattern in the project area was assessed. The farmers were asked about the number of livestock they have and how these animals are fed. It was learnt that the farmers willing to participate in the afforestation and reforestation project activity usually stall fed their cattle from agricultural waste obtained from canal-fed irrigated lands. The goats and sheep are grazed in the lands near railway tracts or on canal-fed agricultural fields after the harvest. It was univocally said that no grazing occurs on the project lands. Migratory cattle also visit these villages, but they graze along the roads and canals. They are not allowed to graze in the degraded croplands. As animal husbandry is also one of the income sources of the farmers, most of them own one or two buffaloes and cows, while a very small percentage owns sheep and goats. It was learnt through interviews that the income from buffaloes is around ₹2000 per annum per buffalo, while it is ₹1000 per annum per cow.
- **Standard of living**: Transect walks, house visits, and direct interviews with family members of the project participants as well as group interviews with the womenfolk of all the eight villages were carried out to study the living conditions of the communities. Apart from a very small percentage of people in the village, it was learnt that all the households lack basic running water and toilet facilities. The lack of various basic amenities was a clear indicative that the farmers participating

in the proposed small-scale afforestation and reforestation CDM project activity belonged to low-income communities.

Summary of Comments Received

The selected areas for the proposed small-scale afforestation and reforestation CDM project activity are severely affected by shifting sand dunes. The land is degraded in nature. The fertility of the lands is poor and not suitable for profitable agricultural ventures. In a few areas, rain-fed agriculture is possible once on an average interval of three to four years. Although a few canal-fed irrigation systems are present in the district of Sirsa, these canals do not pass through the vicinity of the project area.

The climate of the project area is characterized by dryness and can be classified as semi-arid climate. The daily maximum temperature during May–June ranges from 42°C to 47°C. In winter, the area experiences extreme cold and frost.

During the semi-structured interviews, farmers understood the basic concept of CDM and its potential benefits. The villagers agreed that the proposed small-scale afforestation and reforestation CDM project activity could enable them to grow timber trees or fruit trees in degraded lands, which would have otherwise been kept fallow as no other alternative for these lands exists.

Farmers have also understood the conditions of land eligibility for afforestation and reforestation CDM projects and agreed upon that they would contribute only degraded lands (not suitable for profitable agricultural practices) and lands that have not been under forest since the last 50 years. They also agreed to contribute their degraded lands for a total period of 20 years.

The farmers are willing to take part in this project. Their active participation in the discussion and their keen interest in knowing details and the specifications of the project are proof for this (evidences recorded on video).

Farmers who participated in the PRA activity volunteered to give land for the project period of 20 years and were quite happy in being allowed to manage the plantation by them through the society. However, the farmers insisted that each village should be represented in the plantation management committee.

Draft by-laws were elaborately discussed with the farmers (evidence recorded on video). In the brain-storming exercises, it was felt that the farmers are willing to take part in the afforestation activity on their private lands. The PRA team explained in detail the draft by-laws to the farmers for the proposed afforestation activity and also described them various types of memberships, namely, general membership and executive membership.

The society will have a committee of elected members from the general body with representatives from each participating village. The president and the vice president would be elected from the executive body. The secretary of the society would be a serving forest range officer (Ellenabad) who would supervise the proceedings of the society. The society would have full right and control to decide on the proposed small-scale afforestation and reforestation CDM project activity and benefit sharing.

The general members of the society would be the farmers who would take part in the proposed project activity by contributing their degraded fallow marginal agricultural land for afforestation. To broaden the participation by inclusion of women (who do not normally own lands in this district), it has been decided that at least one-third of the executive members should be women.

In the meeting it was also agreed that once contributed for the proposed project, all the lands considered for 20 years would be managed by the landowner under the overall supervision of the society, which would decide further activities on this project area. The farmers agreed to the mechanism of sharing of benefits accruing from the proposed small-scale afforestation and reforestation CDM project activity.

The displacement of activities due to the proposed CDM activity is insignificant. Grazing of livestock is not done in the project lands. Livestock are mostly stall fed in the village. Farmers stated that no open grazing is generally practised in the area, and the cattle are mostly stall fed. Whatever migratory cattle arrive in the area from the adjoining Rajasthan state, they graze in canal-fed fields, which are away from the project area.

The PRA exercise brought out the list of tree species preferred by the farmers for plantation in the project lands. It was agreed upon that a mixed plantation will be carried out, comprising trees with short-, middle-, and long-term rotation. To ensure quick returns, a certain proportion of fruit plants were also selected by the farmers. As per preference, the following trees were listed by the farmers: *Prosopis cineraria, Dalbergia sissoo, Ailanthus excelsa, Acacia nilotica, Acacia tortilis, Azadirachta indica, Tecomella undulata, Melia azadirachta,* and Eucalyptus hybrid.

It is evident from the PRA that farmers were expressing this preference based on the profitability of timber species. But the forest department informed that species such as Dalbergia and Eucalyptus can be successful only in irrigated lands. Availing guidance from the forest department, the farmers took an educated decision about the tree species to be planted in the project lands. They zeroed in on these species: *Ailanthus excelsa, Acacia tortilis,* Eucalyptus hybrid, *Acacia nilotica, Dalbergia sissoo, Prosopis cineraria,* and *Ziziphus mauritiana*. The fruit tree preferred by the farmers was beri.

The stakeholders expect that the proposed small-scale afforestation and reforestation CDM project activity would earn them additional means of income through employment generation and credits earned through sequestration of carbon in their trees.

Land Eligibility of the Project Area

It was evident from various exercises that the farmers own lands that are degraded due to shifting sand dunes. On these lands, rain-fed agriculture is the normal practice. Since the last 50 years, neither the forest department nor the farmers have attempted plantation on these lands. It is mainly because of barriers such as harsh ecological conditions of the area. The productivity of the rain-fed agriculture is also not profitable, considering the amount of resource inputs and returns received.

In the seasonality analysis, the farmers commented that these lands are not utilized for grazing of livestock, so there is no question of any livestock being displaced. Moreover, the lands have less than 5% crown cover (approximately three to four shrubs were observed per 100 m^2 through ocular estimate). The height of these shrubs is 0.5–1.0 m. In the historical analysis, farmers stated that the private lands contributed to the proposed small-scale afforestation and reforestation CDM project activity have been kept fallow since 1970, and only a few agricultural operations were practised depending on the scanty rainfall, which also takes place once in three to four years.

Therefore, this area was not a forest for almost 40–50 years. So these lands satisfy the land eligibility conditions laid down in the methodology for small-scale CDM project activity.

Suitability of Tree Species to be Grown

The PRA exercise brought out the tree species preferred by the farmers for plantation in the project lands. It was agreed upon that a mixed plantation will be carried out comprising trees with short- and medium-term rotation. To ensure quick returns, a certain proportion of fruit plants were also selected by the farmers. As per the preference of the farmers, the following tree species were listed: *Ailanthus excels, Acacia tortilis,* Eucalyptus hybrid, *Acacia nilotica, Dalbergia sissoo, Prosopis cineraria,* and *Ziziphus mauritiana.*

It is evident from the PRA that farmers were expressing this preference based on the profitability of timber species such as *Dalbergia sissoo* and also because of their expectation to get quick return from fruit species such as beri. But with the guidance from the Haryana Forest Department, the farmers took an educated decision about the tree species to be planted in the project lands.

Additionality and Leakage

One of the main objectives of the PRA exercise conducted was to analyse the potentiality of additionality and leakage in the project area. This analysis is important to ensure that the proposed small-scale afforestation and reforestation CDM project activity would lead to positive and additional benefits for the participants and the local communities. Therefore, a few important issues concerning the proposed project's additionality and leakage were discussed during the PRA exercise. The discussions helped the PRA team to understand the present circumstances of the participants and more importantly their requirements, needs, and concerns for livelihood development. The emergent issues of the discussions and comments from the stakeholders are summarized in the following paragraphs.

The land under the project area is degraded by shifting sand dunes since long. The area receives erratic and scanty rainfall (150–200 mm), which leads to water scarcity. The network of canals (Bhakra Nangal canal system and Ghaggar river) in the region does not pass through the project area.

During the summer season, temperature in the project area rises up to 50°C, causing desiccation of plants. Frost is an annual threat to the vegetation in the area, and the region suffered considerable damage in January 2005. Due to high summer temperatures and low rainfall, the area is susceptible to droughts. No plantation activity was attempted by the forest department in the project lands due to these harsh ecological conditions. The above information gathered from the villagers helped to prove that the proposed small-scale afforestation and reforestation CDM project activity will be additional in the area.

The project area is degraded cropland and grassland, which is generally left fallow. Profitable agriculture is carried out on an average interval of three to four years and only one crop is taken instead of two, which is the general trend in the nearby agricultural fields. Thus, the agriculture activity shifting to nearby area is insignificant and far below 50%.

The cattle in the area are stall fed, and the cattle feed comprises agriculture waste as well as fodder crops grown in the adjoining fertile agriculture fields, which are canal irrigated. The region is visited by migratory cattle from Rajasthan in the months of March to May. These cattle feed on agricultural wastes and grasses growing on roadside. Therefore, the number of displaced grazing animals due to the proposed small-scale afforestation and reforestation CDM project activity is unlikely to occur. The above information collected through the PRA exercise indicates that the leakages arising due the proposed project activity would be far below 10%.

Land Condition (Land Eligibility)

- In general, almost all the stakeholders from all the villages informed that there were moderate to dense forest during the 1940s, which rapidly got degraded and finally got lost between 1945 and 1950.
- The rapid loss of forests during this period is attributed to the migration of people during the partition of India and Pakistan, as the area is close to the borders of Pakistan and also to the growth in population of local people with time.
- The forests were cleared manually, basically by migrants and the local people for practising agriculture and for the wood required for constructing houses.

Grazing

- All the stakeholders agreed that almost no grazing is practised in the parcels of land provided for the project.
- This is further confirmed by stakeholders that they stall feed their cattle by providing them fodder from canal-irrigated lands nearby to their villages.

- Regarding migratory cattle from the adjoining state of Rajasthan, they stated that these cattle do not graze in the project area and whatever grazing occurs is on roadside near the canal, away from the project area.

Income

- Stakeholders' consultation as well as field visits and data have clearly shown that the participants belong to low-income groups with an average income ranging from ₹20,000 to ₹40,000 per year.
- The farmers, who have provided their degraded land but also possess tube well-irrigated land, manage to earn livelihoods from them, which ranges from ₹1000 to ₹1500 per acre.

Species Choice

- Species choice by stakeholders was voluntary and based on their interest for anticipated return following a ratio of 4:6 for short- and medium-rotation crops.
- Stakeholders were keen on planting species such as jand and kikar.

Agriculture Potential of Project Lands

- All the stakeholders were unanimous in expressing that the degraded land provided for the proposed small-scale afforestation and reforestation CDM project activity yields a poor crop, that also only after rains once in three to four years.
- The stakeholders also agreed that there is no potential use of this degraded land as it is a matter of loss whenever they have invested money to get crops.

Any Alternatives

- Stakeholders from all the participating villages categorically said that they do not have any other alternative for the degraded land provided for the proposed CDM project activity other than waiting for rains and risking investment to get crops from it after three to four years.

Quotable Comments of the Stakeholders

- "The area has these degraded lands since we were born and they are still the same," a 70 year old stakeholder said.

- "Once dense forest, these areas were cleared by migrants in 1950s."
- "The area had sparse forest, which was completely cleared by the growing local population for wood and fuel."
- "We could hardly get any crop from the degraded land as we have to wait for rains, which occur only once in four to five years."
- "Certainly, we have no other option but to wait for rains to get meagre crop from degraded lands."
- "Most of the time, we have incurred losses by investing in degraded lands for crop, as return has been very poor."
- "The question of cattle grazing in the project area does not arise at all, as there exist not a blade of grass for the animals."
- "Although we earn around ₹30,000 annually from farming and sometimes even more depending on market prices after meeting our daily needs, nothing is left over as saving."

Comments from Local Non-governmental Organizations

Comments from five local non-governmental organizations (NGOs)—Wildlife Society of India; Wildlife Institute of India; Life Line Awareness and Serving Welfare Society; Sirsa, Maksad Welfare Society, Sirsa; and Shikhar Chetna Sangathan—have been collected. All NGOs are convinced and satisfied with all the socio-economic, environmental, and technological aspects of the projects. The specific comments are summarized below.

- The proposed afforestation and reforestation CDM project, being a pilot project in the country, will give impetus to many such project activities in the future. The project will develop the skills of the farming community in deriving multiple benefits from tree growing and crop cultivation. It will also provide an opportunity to the Haryana Forest Department to educate farmers about the need for mitigation of climate change leading to the adoption of the principles of GHG emissions reductions at all levels of society.
- As the first afforestation and reforestation CDM project activity in the region, it is expected to provide excellent opportunities for training and exposure to this fast-developing field and at least some educated young men and women in these villages hope to be able to access entirely new employment avenues in the field of CDM.

- It will also open up new economic opportunities and enhance the income of the farmers through timber and fruit production as well as earning from the sale of carbon credits.
- This project is expected to enhance community bondage and elevate the status of women by ensuring their participation in the proposed afforestation and reforestation CDM project activity.
- The project will not harm the existing flora and fauna as the project area is small and scattered, and the project area only form the fringe area of the natural habitat of the concerned wildlife. Thus, there would be no habitat destruction or fragmentation. On the other hand, plantation established by the proposed project will provide a refuge to some desert wildlife, and the economic benefits of the plantation to the locals will be helpful in establishing a harmonious relationship between them and the conservationists.

F.3 Report on how due account was taken of any comments received:

Due account was taken in incorporating the comments received from the PRA survey in preparation of detailed PRA report, Project Concept Note and PDD documents. The following aspects were especially taken into account:

- The stakeholders were informed about the by-laws and other aspects of the project in public meeting and all their queries answered.
- The participation of local farmers was made on a voluntary basis after they fully understood various aspects of the CDM project, such as anticipated benefits, sharing of usufructs, and so on.
- Local farmers who have contributed their degraded lands voluntarily have unanimously agreed to manage the land themselves under the guidance of the society.
- The preferences of farmers were taken into account in the selection of tree species, which are of economic and ecological importance for the local people.
- Farmers have agreed to use chemical pesticides very sparingly.

Annex 10.1

Contact Information of Participants in the Proposed Small-scale Afforestation and Reforestation CDM Project Activity

Organization: Haryana CDM Variksh Kisan Samiti, Ellenabad, Sirsa
Street/PO Box:
Building: Forest Complex Sirsa
City: Sirsa
State/Region: Haryana
Postfix/ZIP: 125 055
Country: India
Telephone:
FAX:
E-mail: hsamb1998@hry.nic.in
URL:
Represented by: Shree Ravi Khod
Title: President
Salutation: Mr
Last Name: Khod
Middle Name:
First Name: Ravi
Department:
Mobile: +91-9416303222
Direct FAX:
Direct tel: +91-1698-282222
Personal e-mail:

Annex 10.2

Information Regarding Public Funding

No official development assistance will be utilized for the purpose of this proposed small-scale afforestation and reforestation CDM project activity.

Annex 10.3

Declaration on Low-income Communities

Please provide a written declaration that the proposed small-scale afforestation or reforestation project activity under CDM is developed or implemented by low-income communities and individuals as determined by the host party.

There is no single accepted definition of low-income communities by the Government of India for the purpose of CDM projects. Therefore, the international norms have been followed according to which people below a daily income of $1 (about ₹50) fall under the poverty line. Most members (92%) of the society fall in this low-income family category with an average annual income of just ₹6610 as has emerged in the PRA report (refer to E.1).

The main income source of the project participants is farming, and these farmers have small land holdings, a considerable part of which is degraded/degrading and not suitable for profitable agriculture as investment costs are too high for them to carry out any kind of farming or planting activities. Therefore, plantation activities in these lands are possible only through external aid and intervention. The majority of households lack basic amenities such as running water and bathroom facilities. Most farmers reside in small mud houses, which are not concrete structures, and the cattle are also housed in the same dwelling. Fuelwood is the medium of cooking. Most households do not own vehicles for transport. Farmers do not have any savings or assets of any nature apart from the land and a few cattle. The annual income of the household varies from ₹20,000–40,000, which is still far below the international poverty line of ₹80,000.

The above reasons clearly show that the households participating in the proposed small-scale afforestation and reforestation CDM project activity belong to low-income communities.

Appendix 11

LIST OF CARBON OFFSET SUPPLIERS

Name of supplier	Website
AgRefresh	www.agrefresh.org
AMBIO	www.ambio.org.mx
Atrium Carbon Fund LP	www.atriumbrokerage.com
AusCarbon Pty Ltd	www.auscarbongroup.com.au
Australian Carbon Traders	www.australiancarbontraders.com
BalanceCarbon	www.balancecarbon.com
Blue Source LLC	www.bluesource.com
Bonneville Environmental Foundation	www.b-e-f.org
Borealis Offsets	www.borealisoffsets.com
Brighter Planet	www.brighterplanet.com
Camco International	www.camcoglobal.com
CantorCO2e	www.cantor.com
Carbon Balance Consulting	www.carbonbalanceconsulting.com.au
Carbon Clear Ltd	www.carbon-clear.com
Carbon Conservation	www.carbonconservation.com
Carbon Life Ltd	www.carbonlifeltd.com
Carbon Reduction Fund	www.carbonreductionfund.org
Carbonfund.org	www.carbonfund.org
Carbonza P/L	www.carbonza.com
Clean Air Action Crop	www.CleanAirAction.com
Clear, the carbon offset company	www.clear-offset.com
ClearSky Climate Solutions	www.clearskyclimatesolutions.com
Climact	www.climact.com
Climat Mundi	www.climatmundi.fr
Climate Friendly	www.climatefriendly.com

Contd...

Name of supplier	Website
Climate Neutral Group	www.climateneutralgroup.com
CO2logic	www.co2logic.com
CO2OL Forest Finance Group	www.co2ol.de
co2operate BV	www.tiscali.nl
Colorado Carbon Fund	www.coloradocarbonfund.org
Commerce Carbon Exchange	www.commercecarbon.com
Community Energy Inc.	www.newwindenergy.com
Conservation International	www.conservation.org
Cool Planet	www.coolplanet.com.au
CoolClimate LLC	www.atmosclear.org
COzero	www.cozero.com.au
CP Holding LLC	www.carbonlesspromise.com
CPS Carbon Project Solution Inc.	www.carbonprojectsolutions.com
Delta Institute	www.delta-institute.org
Demirer Enerji	www.demirerholding.com
Ducks Unlimited Inc.	www.ducks.org
E.Value—Estudose Projectos de Ambientee Economia, SA	www.evalue.pt
E+Co	www.eandco.net
EcoLogic Development Fund	www.ecologic.org
EcoSecurities	www.ecosecurities.com
Ecosur	www.ecosur.fr
Emergent Ventures Pvt Ltd	www.emergent-ventures.com
Enecore Carbon Ltd	www.enecore.com
Energy Mad Ltd	www.energymad.co.nz
Environmental Capital LLC	www.encapllc.com
Environmental Credit Corp.	www.envcc.com
ERA Ecosystem Restoration Associates Inc.	www.econeutral.com
Essent Trading International SA	www.essenttrading.com
Evolution Markets	www.evomarkets.com
First Climate	www.firstclimate.com
Forests NSW	www.sf.nsw.gov.au
Global Green Energy LLC	www.globalgreen-energy.com
global woods AG	www.global-woods.com
GoodPlanet/Action Carbone	www.actioncarbone.org

Contd...

Name of supplier	Website
Green Carbon Ltd	www.greencarbon.co.nz
Green Mountain Energy	www.greenmountain.com
Greenhouse Balanced	www.greenhousebalanced.com.au
Greenhouse Gas Services	www.ghgs.com
Grupo Ecológico Sierra Gorda	www.prodigy.net.mx
Habitat Enterprises Ltd	www.habitatenterprises.ca
J.P. Morgan Climate Care	www.jpmorganclimatecare.com
JJCENG.COM, PC	www.jjceng.com
LiveCooler Foundation, Inc.	www.livecooler.org
LMS Generation Pty Ltd	www.lms.com.au
Low Energy Supplies and Services Pty Ltd	www.lowenergy.com.au
Meridian Energy	www.meridianenergy.co.nz
MF Global	www.mfglobalenergy.com
Myclimate Foundation	www.myclimate.org
NativeEnergy, Inc.	www.nativeenergy.com
North Coast Forest Conservation Programme	http://conservationfund.org
Offsetters	www.offsetters.ca
OneCarbon	www.onecarbon.com
Origin Energy	www.originenergy.com.au
Pacific Forest Trust	www.pacificforest.org
Paso Pacífico	www.pasopacifico.org
Prime Carbon Pty Ltd	www.primecarbon.com.au
SKG SANGHA	www.skgsangha.org/
South Pole Carbon Asset Management	www.southpolecarbon.com
Southern Metropolitan Regional Council	www.smrc.com.au
Standard Carbon LLC	www.standardcarbon.com
Sterling Planet, Inc.	www.sterlingplanet.com
Terra Commodities Limitada	www.terrabro.com
TerraPass Inc.	www.terrapass.com
TFS Green	www.tfsgreen.com
The CarbonNeutral Company	www.carbonneutral.com
The Climate Trust	www.climatetrust.org
The Conservation Fund GoZero	www.conservationfund.org
The Nature Conservancy	www.tnc.org

Contd...

Name of supplier	Website
The PACE Centre	www.carbon.org.za
The Trust for Public Land	www.tpl.org
The Woodland Trust	www.woodlandtrust.org.uk
Trees for Travel	www.treesfortravel.nl
Trees, Water & People	www.treeswaterpeople.org
Tricorona	www.tricorona.se
Tullett Prebon	www.tullettprebon.com
Village Green Energy	www.villagegreenenergy.com
Voltalia	www.voltalia.com
World Land Trust	www.worldlandtrust.org
Zero Emissions Technologies, SA	www.zeroemissions.abengoa.com
Zerofootprint	www.zerofootprint.net
ZeroGHG Inc.	www.zeroghg.com

Appendix 12

LOCATION OF SOME IMPORTANT CARBON FORESTRY PROJECTS IN INDIA, WITH SALIENT FEATURES*

HIMACHAL PRADESH REFORESTATION PROJECT (1)

Overview

The project has been developed through a series of consultations with Mid-Himalayan Watershed Development Project (MHWDP) and its stakeholder constituents, namely, Forest Department, Government of Himachal Pradesh, local Gram Panchayats (GPs), and the World Bank. The project seeks to implement afforestation and reforestation CDM activities on 10,000 ha of degraded lands in the watersheds of mid-

* Details available at <www.forestcarbonportal.com/project/international-small-group-and-tree-planting-programme-list>

Himalayan region. The four guiding principles of the project are as follows:

1. Adoption of native and locally preferred tree species for reforestation.
2. Involvement of the local GPs and small and marginal farmers in reforestation activities, which will strengthen the ongoing watershed interventions.
3. Facilitation of technical, financial, and capacity development support from MHWDP to reforestation activities.
4. Distribution of carbon revenue to the village community (GPs and farmers).

The project is expected to bring value addition to the ongoing physical catchment/drainage treatment activities undertaken as part of the MHWDP. It seeks to restore degraded lands through reforestation activities. The three plantation forestry models considered for the purpose are as follows:

1. **Restoration forestry model:** In this model, reforestation of degraded non-classed or non-demarcated/demarcated protected forestland is proposed with a tree density of 1100 plants per hectare. Largely native species would be planted under this model. This model aims to protect the watersheds and regeneration of native flora, supplemented with planting of native tree species on degraded sloping high altitude lands in selected GPs. This is also projected to lead to conservation of biodiversity. Regenerated forests are expected to provide non-timber forest products to local communities and improve their livelihood opportunities. This model is proposed to cover 7095 ha.
2. **Community forestry model:** This model is proposed for reforestation of degraded community land (common land). Largely native species are included in this model. The reforestation activity will lead to protection of watersheds and improvement of biomass required to meet the community needs such as small timber, fuelwood (woody litter), fodder collection for livestock, and harvest of non-timber forest products. This model is expected to cover about 775 ha, and the density of planting in this model is also 1100 trees per hectare.
3. **Farm forestry model:** This model covers an area of 2130 ha and includes reforestation of abandoned or long-term fallow private land with tree species largely providing fruits and fodder to the landowners. The density of planting is 1100 trees per hectare. Landowners will

also derive fuelwood from fallen woody litter. This model will provide employment to the landowners apart from protecting the abandoned hilly land.

The project will be implemented by the MHWDP, which is an administrative unit of the Forest Department of Himachal Pradesh. Many of the project activities such as protection and management would involve participation of local panchayats and their delegated committees.

The afforestation and reforestation CDM project is developed under the umbrella of the World Bank-funded MHWDP, and it is expected to sequester 5,058,103 tonnes of CO_2 equivalent of tCERs over the first crediting period of 20 years at the rate of 25.29 tonnes of CO_2 equivalent per hectare per year.

Location: Bilaspur, Himachal Pradesh, India

Project Developer(s): MHWDP will manage the project till 2013 or beyond (if extended) and subsequently the State Forest Department will manage the project.

URL: http://cdm.unfccc.int/Projects/Validation/DB/ZUOWR7RIVG6BECVHP1E229QEN3GMQK/view...

Project Start Date: 2006

Hectares: 10,000

Standard(s): CDM

Verifier(s): TUV-SUD (CDM Validation)

Total Reductions: 5,058,103

Crediting Period: 20 years

Credit Buyers: Spain (BioCarbon Fund)

Additional Information: CDM ID—CDM5022

Source <www.forestcarbonportal.com/project/international-small-group-and-tree-planting-program-list>

THE INTERNATIONAL SMALL GROUP AND TREE PLANTING PROGRAM (TIST) (2)

Overview

Since its inception in 1999, over 40,000 participants organized into more than 5600 TIST small groups have planted over 6 million trees in Tanzania, India, Kenya, and Uganda, accomplishing GHG sequestration through tree planting, creating a potential long-term income stream, and developing sustainable environments and livelihoods. Replication of TIST in India began in June 2002. Currently, over 5000 TIST participants in more than 800 small groups are registered in the TIST programme in Northern Tamil Nadu, India, and are working to break their local cycle of deforestation, drought, and famine. The trees have already begun to reduce erosion, stabilize and enrich the soil, and will soon provide shade. In the future, they will provide other benefits, including edible fruits and nuts, medicines, windbreaks, firewood, and timber.

This PDD is for a reforestation project and applies to 111 of the small groups, 1200 members, 175 project areas, and 106 ha. The main species planted are *Casuarina equisetifolia, Eucalyptus grandis,* and *Tectona grandis.*

As a grassroots initiative, small groups are provided with a structural network of training and communications that allows them to build on their own internal strengths and develop best practices. Small groups benefit from a new income source; the sale of carbon credits that result from the sequestration of carbon from the atmosphere in the biomass of the trees and soil. These credits are expected to be approved for CDM and, because they are tied to tree growth, will be sustainable. Thus, the carbon credits create a new opportunity for sustainable development to the beneficiaries.

Location: Kancheepuram, Tamil Nadu, India

Project Developer(s): Tree Planting India Pvt. Ltd

Project Start Date: 2004

Hectares: 106

Species Planting Mix: Native and exotic plantings (more than 50% native)

Standards: CDM

Verifier(s): TUV-SUD (CDM Validation)

Total Reductions: 107,810

Crediting Period: 30 years

Credit Buyers: United Kingdom (Climate Change Capital)

Additional Information: CDM ID—CDM3776

SMALL-SCALE AFFORESTATION CDM ON PRIVATE LANDS (3)

Overview

Small-scale Cooperative Afforestation CDM Pilot Project Activity on Private Lands Affected by Shifting Sand Dunes in Sirsa, Haryana

The lands to be planted in the proposed small-scale afforestation and reforestation CDM project activity are located in the western belt of Haryana, which has its border with the state of Rajasthan at the north-eastern fringe of the Indian Thar Desert. The project area is affected by aeolian (wind-blown) sand, and it is the degraded part of croplands spread across these eight villages, comprising 369.87 ha and belonging to 227 farmers, which is generally left fallow. Large areas of land are without any vegetation due to frequent dust storms of various intensities. These dust storms toss up large amount of sand, dust, and suspended particles into the air and pollute the ambient atmosphere. The report has found that the quality of drinking water and the water table in this region have deteriorated over the years. Many villages also reportedly have lost croplands due to shifting sands. Impacted by limited precipitation (100–200 mm annually) and shifting sand dune, the cropping intensity on these degraded croplands is barely one crop every three years as against the normally two crops annually on the surrounding good croplands (as per the PRA findings). The cultivation and shifting sand dunes prevent the potential natural regeneration of forest in this area. The objectives of the small-scale afforestation and reforestation CDM project activity proposed by Haryana CDM Variksh Kisan Samiti, Ellenabad, Sirsa (Haryana CDM Tree Farmers Society) are as follows:

- To earn carbon credits from growing of trees under the CDM provisions of the Kyoto Protocol.
- To help in mitigation of global warming by planting trees for sequestration of atmospheric carbon dioxide.
- To improve the local environmental condition of soil through increasing the water-holding capacity of the lands, increasing the humus in soil, and also stabilizing the sand dunes, by converting the marginal and degraded croplands into forested lands.
- To increase income, provide employment opportunities, and as a result to alleviate poverty of local communities.

To realize the objectives mentioned above, 369.87 ha of mixed forests will be established, using seven tree species: *Ailanthus excelsa, Acacia tortilis,* Eucalyptus hybrid, *Acacia nilotica, Dalbergia sissoo, Zizyphus mauritiana,* and *Prosopis cineraria.*

The proposed small-scale afforestation and reforestation CDM project activity is a pilot project activity of its kind in Haryana. Both the project developer (Haryana Forest Department) and the local farmers expect that the success of the proposed project activity will promote afforestation and reforestation activities in lands of low agricultural productivity in Haryana and beyond. They also believe that it will contribute to poverty alleviation, biodiversity conservation, and prevention of soil erosion, thus contributing to sustainable development.

Location: Haryana, India

Project Developer(s): Haryana CDM Variksh Kisan Samiti, Ellenabad, Sirsa

URL: http://cdm.unfccc.int/UserManagement/FileStorage/DJISYMVG9RKHTCPU53X4BZA0NW8O2L

Hectares: 370

Standards: Clean Development Mechanism (CDM)

Credit Status: Other

Market: CDM

Total Reductions: 231,920

Crediting Period: 20 years

Registry: CDM

Additional Information: hsamb1998@hry.nic.in

REFORESTATION OF SEVERELY DEGRADED LANDMASS—ITC SOCIAL FORESTRY (4)

Overview

The Society for National Integration through Rural Development (SNIRD) and the Society for Elimination of Rural Poverty (SERP), both promoted by the Government of Andhra Pradesh, are the bodies that have identified the tribal beneficiaries and grouped them into a Sangha (user groups/society) for taking up the plantation activity. Apart from providing finances for the project, ITC also distributes planting stock nurtured from hybrid clonal plants of Eucalyptus developed at ITC's own research and development centre at Bhadrachalam. ITC continues to extend services such as sharing of agricultural/forestry knowledge and good practices of farming to the beneficiaries identified under the project activity. The following are the purposes of the proposed project activity:

- To create a long-term secure income for the rural poor (tribals).
- To sequester CO_2 through reforestation activities.
- To improve soil and water erosion control measures in the Godavari catchment area.

The project activity has been able to create carbon sink of 3070.19 hectares, which not only helps in sequestration of CO_2 but also has several other beneficial effects that have been delineated under the section below on Sustainability Development by the Project Activity. The project activity has also been able to create appreciable impression among the farmers in the region who have approached ITC for assistance with good farming practices and help them regenerate their degraded land.

The species considered for carbon sequestration are *Eucalyptus tereticornis* (Smith) and *Eucalyptus camaldulensis* (Dhen), which belong to the family Myrtaceae of Angiospermae. Even though the species selected are exotic, introduced from Australia, they have been grown throughout the length and breadth of India for the past 200 years. The species have not exhibited any invasive behaviour, as natural regeneration is absent. These are also not considered genetically modified organisms (GMOs).

The Ministry of Environment and Forests (MoEF) of the Government of India has stipulated the social, economic, environmental, and technological well-being as the four indicators for sustainable development in the interim approval guidelines for CDM projects.

Location: Andhra Pradesh, India

Project Developer(s): ITC Ltd, Paperboards and Specialty Papers Division (PSPD), Bhadrachalam Unit

URL: http://cdm.unfccc.int/Projects/DB/BVQI1222275709.04/view

Project Start Date: 2001

Hectares: 3070

Standards: CDM

Market: Voluntary Over-the-Counter Market

Total Reductions: 1,733,753

Crediting Period: 30 years

Additional Information: Pradeep Dhobale (pradeep.dhobale@itc.in)

IMPROVING RURAL LIVELIHOODS THROUGH CARBON SEQUESTRATION BY ADOPTING ENVIRONMENT FRIENDLY TECHNOLOGY-BASED AGROFORESTRY PRACTICES (5)

Overview

The proposed afforestation and reforestation CDM project activity will mobilize resource-poor farmers to raise tree plantations on farmlands. It proposes to link resource-poor farmers and end users of wood products to optimize the land use and to facilitate the coordination of wood producers, agronomists, financial institutions, and NGOs to improve the livelihood opportunities of rural households. The project activity is implemented on degraded farmlands or lands used for rain-fed subsistence agriculture.

The project is implemented in the two states of India: Odisha and Andhra Pradesh. The project area includes small landholders spread over a total of six districts: Rayagada, Koraput, and Kalahandi in Odisha and Visakhapatnam, Srikakulam, and Vizianagaram in Andhra Pradesh. These districts have a predominance of indigenous population, notified as Scheduled Tribes and Scheduled Castes in India, with the majority of them being poor.

Odisha is one of the poorest states in India in terms of per capita income. In Odisha, about 41% of population belongs to Scheduled Castes and Scheduled Tribes. Odisha also has the largest number of indigenous communities or tribes in the country, that is, 62 tribes constitute about 23% of the total population. The tribal population is predominant in Koraput, Rayagada, and Kalahandi districts.

In Andhra Pradesh, the project area covers 33 indigenous communities or tribes in the three districts of Srikakulam, Vishakapatnam, and Vizianagaram. The major tribes represented in the project area include Andh, Bagata, Bhil, Chenchu, Chenchuar, Gadabas, and Gond.

The participation of small and marginal farmers representing the indigenous communities and their organization as part of the CDM activity makes this project unique in contributing to their land-use choice, improvement of livelihood opportunities, and in promoting their capacity to organize and implement climate change mitigation initiatives.

The project implements reforestation on 3607.32 ha of degraded lands in Andhra Pradesh and Odisha. The reforestation activities under the project were initiated in 2004 with a plan to complete them over a period of four years.

Location: Koraput, Odisha, India

Project Developer(s): VEDA Climate Change Solutions Ltd; JK Paper Ltd

URL: http://cdm.unfccc.int/Projects/Validation/DB/18OZONIVT66PIKEQ6APYNE8RHCUCMX/view...

Project Start Date: 2004

Hectares: 3607.32

Standards: CDM

Verifier(s): TUV-SUD (CDM Validation)

Total Reductions: 312,137

Crediting Period: 30 years

Credit Buyers: Canada (BioCarbon Fund)

Additional Information: CDM ID—CDM4993

REFORESTATION OF SEVERELY DEGRADED LANDMASS IN KHAMMAM DISTRICT OF ANDHRA PRADESH, INDIA, UNDER ITC SOCIAL FORESTRY PROJECT (6)

Overview

Under the project activity, the degraded lands owned by the rural poor (tribals) are developed for raising plantations with Eucalyptus. The Paperboards and Specialty Papers Division (PSPD) is the primary developer of this project, and it initiated the project activity through the local NGOs. The participating NGOs include Action for Collective Tribal Improvement and Vocational Education (ACTIVE), Homeless Outreach Poverty Eradication (HOPE), Society for Health and Agriculture Department (SHADE), SNIRD, and SERP, all promoted by the Government of Andhra Pradesh. These bodies identified the tribal beneficiaries and grouped them into a Sangha (user group/society) for taking up the plantation activity. Apart from providing finances for the project, ITC also distributes planting stock nurtured from hybrid clonal plants of Eucalyptus developed at ITC's own research and development centre at Bhadrachalam. ITC continues to extend services such as sharing of agricultural/forestry knowledge and good practices of farming to the beneficiaries identified under the project activity.

The project activity has been able to create carbon sink of 3070.19 ha, which not only helps in sequestration of CO_2 but also has several other beneficial effects. The project activity has also been able to create appreciable impression among the farmers in the region who have approached ITC for assistance with good farming practices and help them regenerate their degraded land.

The species considered for carbon sequestration are *Eucalyptus tereticornis* (Smith) and *Eucalyptus camaldulensis* (Dhen), which belong to the family Myrtaceae of Angiospermae. Even though the species selected are exotic, introduced from Australia, they have grown throughout the length and breadth of India for the past 200 years. The species have not exhibited any invasive behaviour, as natural regeneration is absent. These are also not considered GMOs.

The MoEF, Government of India, has stipulated the social, economic, environmental, and technological well-being as the four indicators for sustainable development in the interim approval guidelines for CDM projects. The view of the participant on the contribution of the proposed afforestation and reforestation CDM project to sustainable development is aligned with the sustainable development criteria of the MoEF.

Location: Khammam, Andhra Pradesh, India

Project Developer(s): Maderas Preciosas Nicaragua SA; Precious Woods Holding Ltd

URL: http://cdm.unfccc.int/Projects/DB/BVQI1222275709.04

Project Start Date: 2001

Hectares: 3070.19

Species Planting Mix: Native and exotic plantings (more than 50% native)

Standards: CDM

Verifier(s): BV Cert (CDM Validation)

Crediting Period: 30 years

Additional Information: CDM ID—CDM0585

REFORESTATION OF DEGRADED LAND IN CHHATTISGARH, INDIA (7)

Project Status: Operational

Project Developer(s): Prakash Industries Ltd

Contact Details:

Organization: Prakash Industries Ltd, Street/PO Box: Ring Road No. 2, New Industrial Area, Gogaon, Raipur

State/Region: Chhattisgarh

Postfix/ZIP: 493221

Country: India

Telephone: 91-771-2327598

FAX: 91-771-2327588

URL: www.prakash.com

Department: Plantations and Development

Mobile: 91-9826902468; **Direct FAX**: 91-771-2327588; **Direct tel**: 91-771-2327598

Personal e-mail: manoj0664@yahoo.com

Location: Chhattisgarh, India

URL: www.climate-standards.org/projects/files/chhattisgarh/Prakash_Industries_

Project Type: Afforestation or Reforestation

Project Start Date: 2007

Standards: Climate, Community and Biodiversity (CCB)

Methodology: CDM AR-AM0001: Reforestation of degraded land, v.3

Verifier(s): Rainforest Alliance

Appendix 13

A Run-through Example of Developing an Afforestation and Reforestation CDM PDD

This appendix describes, by way of an example, how an afforestation and reforestation Clean Development Mechanism (CDM) project activity design document can be developed. Even though the example is set in a hypothetical situation, it illustrates practical application of the rules of the CDM. We select the approved methodology AR-ACM003 for our project because we do not have any wetland in our project area.

SCENE SETTING

Imagine we are working in a developing country's forest reserve. Among extensive barren tracts of forest land, we find a small watershed of 300 ha, which has been lying barren[1] for several decades. We plan to consider this area for our afforestation and reforestation CDM project. Within the area, however, there are a few groves of trees preserved by people as sacred groves. Of these groves, three are larger than 0.05 ha and hence must be excluded from the project boundary because 0.05 ha is the minimum area threshold reported by the Designated National Authority (DNA) of the host country for defining forest for the purpose of CDM. Tree groves occupying less than 0.05 ha do not constitute forest, even if these have a crown cover exceeding the minimum tree crown cover threshold. These large groves are surveyed in field, and their location and shape are carefully plotted on the map.[2]

---

1 By "barren" we mean that the tree crown cover, on an average, is less than 15%, which is the threshold tree crown cover for defining forest for the purpose of hosting afforestation and reforestation CDM projects in the host country.

2 Note that there is no requirement for putting up a physical boundary or a fence around these groves. Also, if a part of the watershed were to have numerous small tree groves each occupying an area greater than 0.05 ha, then we could have excluded the whole block of area in which these groves were located if we did not want to survey individual groves because of the cost involved.

We find that the total area occupied by these large groves is 4 ha. Thus, out of 300 ha, an area of 296 ha is eligible for our project.

LAND ELIGIBILITY

The land, excluding the area occupied by the three large tree groves, is eligible for implementing an afforestation and reforestation CDM project activity because it had no forest as on 31 December 1989.[3] We provide the evidence of this fact by producing two satellite imageries of the area—one dating before 31 December 1989, and the other dating after this date.[4]

BASELINE SCENARIO

To determine the baseline scenario for this land, we examine the possible uses of this land in the coming years. Since the land is reserved for forests, no other activity is expected to take place here except reforestation. Is reforestation likely to happen here? We note that reforestation in this area is not expected to take place in the foreseeable future because no private investment can take place in this land (which is state-owned); and the provisions in the public budget (either in the national budget or in the provincial budget) are not likely to lead to reforestation. This is sufficient evidence to prove that the most likely baseline scenario for this land is continuation of the status quo (that is, the land will continue to lie barren as it is today).

BASELINE REMOVALS

The next step is to estimate the changes in carbon stocks in the carbon pools of this land that would occur in the baseline scenario. The land contains a few scattered (and deformed) trees per hectare and very scanty woody shrub vegetation. The trees are not likely to put on any increment since occasional fuel gathering activity, along with grazing by animals during post-monsoon months, takes place here. The level of biomass per hectare is more or less static over the last several years, or

---

3 This requirement comes from the tool "Demonstration of eligibility of lands for afforestation and reforestation CDM project activities".

4 However, we must note that satellite imageries are not the only acceptable evidence for this purpose. We could also have met this requirement by conducting a participatory rural appraisal (PRA) exercise and producing the outcome as an evidence of the fact that there was no forest in the area on that date. The tool on eligibility of land allows us several other options as well.

it is slightly declining. Therefore, it is safe to assume that the expected net increase in the above-ground biomass in the area is zero.[5]

PRE-PROJECT CARBON STOCKS

Our next step is to estimate how much tree biomass per hectare is present in the area at the start of the project.[6] If we determine the precise value of this parameter (that is, tree biomass per hectare), it might require a lot of work and turn out to be costly in comparison to its importance. So we decide to get a "conservative" estimate of this parameter. To do this in a simple way, we estimate the upper bound of the mean tree crown cover in the area. For this, we randomly select 20 plot centres spread over the 296 ha area and at each plot centre we drive a peg into the ground and attach a 30 m tape to the peg and swing the free end of the tape into a circle and count all the trees present within this circular plot. We also measure approximately the crown diameters of the trees present within the plot. Some plots may have no tree at all; others may have a few trees. For each plot, we calculate the percentage tree crown cover by dividing the total crown area of trees by the area of the plot (which is $\varpi \times 30^2$ or 2827 m^2). On summarizing plot data (see Table A13.1), we find that the average tree crown cover in the area is 2.56% at the most, that is, the upper bound of the crown cover is 2.56%. We multiply this by the parameter "above-ground biomass content" from Table 3A.1.4 of the *IPCC Good Practice Guidance for Land Use, Land-Use Change and Forestry* (IPCCGPG-LULUCF 2003)[7] and

5 However, we must note that even if the tree biomass in the baseline were to be increasing, our afforestation and reforestation CDM project could still have gone ahead. We could have used the default method for estimation of tree biomass increment as provided in the tool "Estimation of carbon stocks and change in carbon stocks of trees and shrubs in afforestation and reforestation CDM project activities" using the simplified "proportionate crown cover" method.

6 Note that "carbon stock in tree biomass" is different from the "change in the carbon stock in tree biomass". The fact that one is zero does not mean that the other is also zero. The pre-project carbon stock in tree biomass is estimated and provided in the PDD, but it enters into calculations only at the time of the first monitoring; it must be subtracted from the project tree biomass estimated from sample plot measurements at the time of monitoring because the pre-project trees will get measured along with the project trees. However, if we could avoid these trees from being counted as project trees, which would make our inventory more complicated, we could also exclude their biomass from accounting.

7 This is provided in the simplified "proportionate cover method" available in the tool "Estimation of carbon stocks and change in carbon stocks of trees and shrubs in afforestation and reforestation CDM project activities". Since the default method will not significantly affect our final estimation of tCERs (because the biomass in the baseline of our project is very small), we opt for the default method of estimation. If our baseline were to have a large crown cover (say 12%), we would have considered measuring biomass in sample plots, either by harvesting and weighing all woody vegetation in smaller plots or by using tree allometric equations in conjunction with DBH measurements of all trees in each plot.

we find that the above-ground tree biomass in the area is 1.869 tonnes dry matter per hectare.[8]

Table A13.1 Plot values of tree crown cover

Plot number	Number of trees	Mean crown diameter (m)	Plot crown area (m^2)	Plot crown cover (m^2)
1	3	2.3	12.46	0.44%
2	2	3.1	15.10	0.53%
3	0	0	0.00	0.00%
4	0	0	0.00	0.00%
5	1	3.6	10.18	0.36%
6	5	2.8	30.79	1.09%
7	3	3.4	27.24	0.96%
8	8	2.6	42.47	1.50%
9	0	0	0.00	0.00%
10	0	0	0.00	0.00%
11	11	2.1	38.10	1.35%
12	6	4.5	95.43	3.38%
13	24	3.6	244.29	8.64%
14	19	2.4	85.95	3.04%
15	7	1.9	19.85	0.70%
16	8	2.6	42.47	1.50%
17	21	3.1	158.50	5.61%
18	0	0	0.00	0.00%
19	22	2.7	125.96	4.46%
20	0	0	0.00	0.00%
SD				2.29%
SEM				0.51%
$t_{90\%}$				1.729%
$CI_{90\%}$				0.89%
LB				0.79%
UB				2.56%

[8] We find that "above-ground biomass content (dry matter) in forest" in the host country is listed as 73 tonnes/ha in the IPCC table mentioned in the tool.

We increase this by 25% to account for the root biomass[9] of the trees and find that pre-project tree biomass in the area is 2.336 tonnes dry matter per hectare, which is equivalent to a carbon stock[10] of 4.283 tonnes of CO_2 equivalent per hectare. We note that shrub crown cover in the area is far below 5%, and therefore we do not need to estimate carbon stock in the pre-project shrub biomass.[11]

We further note that since carbon stock in tree biomass in the baseline is not likely to increase, the carbon stock in the carbon pools of dead wood and litter is also not likely to increase. We, therefore, account baseline changes in carbon stock in the above-ground biomass, below-ground biomass, dead wood, and litter pools as zero.

EX-ANTE ESTIMATION OF ACTUAL NET GHG REMOVALS

One of the requirements in the PDD of an afforestation and reforestation CDM project is to estimate in advance how much greenhouse gas (GHG) removals are likely to be achieved under the project. This is done by projecting the growth of the trees to be planted or to be regenerated. In our project, we propose an assisted natural regeneration method: cutback operations combined with seed sowing on contour trenches and plantation of nursery-raised seedlings in gaps. We will use native species only, since our aim is to restore the forest to its natural state. For projecting the biomass growth of a mixed forest, we cannot use any species-specific increment data (and, what is more, no such data are available). Therefore, we estimate annual diameter increments using data from sample trees existing in the project area (or a nearby area) and use a default allometric equation[12] to convert tree diameter into tree biomass.

---

[9] A default value of 0.25 for the root–shoot ratio of trees in the baseline is provided in the tool.

[10] To convert a given quantity of tonne dry matter (tdm) of tree biomass into carbon stock, we multiply it first by 0.47 (which is the carbon content of tree biomass, called carbon fraction), and then by 44/12 (which is the ratio of molecular weights of carbon dioxide and carbon).

[11] Note that if the shrub crown cover were to be somewhere near 5%, we would be required to estimate it. The same sample plots that were used for estimation of tree crown cover could have been used for estimation of shrub crown cover.

[12] The equation selected by us is $AGBM = \exp[-1.996+2.32*\ln(D)]$, which we take from "Estimating Biomass and Biomass Change of Tropical Forests: a Primer. (FAO Forestry Paper - 134)" by S Brown. We select the equation for dry forests growing in an area receiving annual rainfall of more than 900 mm, which is the case for our project area. We must note, however, that ex-ante estimation of tree biomass is only an approximate estimation and no precision requirements are prescribed by the methodology for this. Therefore, we could also have used any other allometric equation (for example, an equation from Table 4A.1 of IPCC-GPG-LULUCF 2003).

To obtain a weighted average of expected diameter increments in dominant tree species in the area, we note that 80% of trees in the area are covered by five tree species, although in different proportions. We can, therefore, estimate the average expected diameter increment on the basis of these five tree species.

To get annual diameter increments expected in trees of these species, we measure DBH of three sample trees of each species and determine their age by counting growth rings after felling the trees. The data obtained are summarized in Table A13.2, and they show that the mean annual increment in DBH, averaged over the five-tree species, is 1.09 cm per year.

Table A13.2 Mean diameter increment observed in sample trees

	Tree 1		**Tree 2**		**Tree 3**		**Mean**		**Diameter increment (cm/year)**		
Species	*Diameter (cm)*	*Age (year)*	*Diameter (cm)*	*Age (year)*	*Diam eter (cm)*	*Age (year)*	*Diam eter (cm)*	*Age (year)*	*Spec ies mean*	*spec ies weight*	*Contr.*
A	14	15	17	20	34	26	21.67	20.33	1.07	0.50	0.53
B	18	20	14	16	16	15	16.00	17.00	0.94	0.20	0.19
C	21	14	15	15	11	15	15.67	14.67	1.07	0.10	0.11
D	22	13	25	16	17	15	21.33	14.67	1.45	0.10	0.15
E	8	7	11	9	9	8	9.33	8.00	1.17	0.10	0.12
										1.00	1.09

However, we know that tree diameter increments are not constant over the lifespan of trees. The diameter increments will be smaller in the beginning, higher in mid-age years, and decrease towards maturity. Therefore, we transform the DBH increments into a symmetric triangular (linear) distribution over a period of 30 years, that is, we assume that for the first 15 years, the DBH increment will be uniformly increasing, and during the last 15 years, the DBH increment will be uniformly decreasing,[13] while the mean increment during the 30-year period will be

[13] We do this for reasons of simplicity. We could also have used a logistic growth function to project tree diameter over time. The alternative values shown in columns (d) and (f) of Table A.3 were computed using the logistic growth equation $D_t = \phi_1/\{1 + \exp[-(t - \phi_2)/\phi_3]\}$, where ϕ_1 is the diameter at maturity, ϕ_2 is the time required to attain half the diameter at maturity, and ϕ_3 is the time elapsed between the two events of attaining (1) half the diameter at maturity, and (2) three-fourths of the diameter at maturity.

1.09 cm. The resulting diameter increments and the expected diameters of the mean tree are shown in columns (c) and (e), respectively, of Table A13.3.

Table A13.3 Tree diameter increment over project period (cm)

Year	ΔD_{CONST}	ΔD_{TRIANG}	ΔD_{LOGIS}	D_{TRIANG}	D_{LOGIS}
(a)	(b)	(c)	(d)	(e)	(f)
1	1.09	0.07	0.00	0.07	0.00
2	1.09	0.22	0.50	0.29	0.50
3	1.09	0.36	0.58	0.65	1.08
4	1.09	0.51	0.66	1.16	1.73
5	1.09	0.65	0.75	1.82	2.48
6	1.09	0.80	0.84	2.62	3.32
7	1.09	0.94	0.94	3.56	4.26
8	1.09	1.09	1.05	4.65	5.31
9	1.09	1.24	1.15	5.88	6.46
10	1.09	1.38	1.26	7.27	7.72
11	1.09	1.53	1.36	8.79	9.08
12	1.09	1.67	1.45	10.46	10.54
13	1.09	1.82	1.53	12.28	12.07
14	1.09	1.96	1.60	14.24	13.66
15	1.09	2.11	1.64	16.35	15.30
16	1.09	2.11	1.66	18.45	16.97
17	1.09	1.96	1.66	20.42	18.63
18	1.09	1.82	1.64	22.23	20.27
19	1.09	1.67	1.60	23.90	21.86
20	1.09	1.53	1.53	25.43	23.40
21	1.09	1.38	1.45	26.81	24.85
22	1.09	1.24	1.36	28.05	26.21
23	1.09	1.09	1.26	29.14	27.47
24	1.09	0.94	1.15	30.08	28.62
25	1.09	0.80	1.05	30.88	29.67
26	1.09	0.65	0.94	31.53	30.61
27	1.09	0.51	0.84	32.04	31.45
28	1.09	0.36	0.75	32.40	32.20
29	1.09	0.22	0.66	32.62	32.85
30	1.09	0.08	0.58	32.70	33.43

Putting the expected tree diameters into the allometric equation, we estimate the expected above-ground biomass of each tree. We expect to have a maximum stocking density of 1250 trees per hectare in the beginning, which will stabilize over time to a stocking density of 550 trees per hectare. Combining these values we calculate above-ground tree biomass per hectare over the 30 years of our proposed project. We expand the above-ground tree biomass to total tree biomass by using the default root–shoot ratios computed from the equation provided in the tool.[14] Finally, we convert tree biomass into equivalent carbon stocks. The results of our calculation are summarized in Table A13.4.

Table A13.4 Ex-ante estimation of carbon stock in tree biomass per hectare

Project year	Mean DBH (cm)	Biomass (kg/tree)	Stocking (tree/ha)	Average tree biomass (t/ha)	Root–shoot ratio	Tree biomass (t/ha)	Carbon stock in tree biomass (tCO_2e/ha)
1	0.07	—	800	—	—	—	—
2	0.29	—	1100	—	—	—	—
3	0.65	—	1250	—	—	—	—
4	1.16	—	1000	—	—	—	—
5	1.82	—	900	—	—	—	—
6	2.62	—	800	—	—	—	—
7	3.56	—	700	—	—	—	—
8	4.65	4.80	600	2.88	0.312	3.78	6.93
9	5.88	8.30	575	4.77	0.301	6.21	11.38
10	7.27	13.53	550	7.44	0.291	9.61	17.61
11	8.79	21.05	550	11.58	0.282	14.84	27.21
12	10.46	31.53	550	17.34	0.273	22.08	40.48
13	12.28	45.71	550	25.14	0.266	31.82	58.34
14	14.24	64.47	550	35.46	0.259	44.64	81.85
15	16.35	88.79	550	48.83	0.253	61.19	112.18
16	18.45	117.63	550	64.70	0.248	80.73	148.00
17	20.42	148.70	550	81.79	0.243	101.70	186.45
18	22.23	181.21	550	99.67	0.240	123.58	226.57

[14] The equation regressing root–shoot ratio on above-ground tree biomass content is: $R = \exp[-1.085 + 0.9256*\ln(b)]/b$, where b is the above-ground tree biomass in tonnes of dry matter per hectare.

Table A13.4 *Contd...*

Project year	Mean DBH (cm)	Biomass (kg/tree)	Stocking (tree/ha)	Average tree biomass (t/ha)	Root–shoot ratio	Tree biomass (t/ha)	Carbon stock in tree biomass (tCO_2e/ha)
19	23.90	214.40	550	117.92	0.237	145.86	267.41
20	25.43	247.49	550	136.12	0.234	168.03	308.06
21	26.81	279.78	550	153.88	0.232	189.63	347.65
22	28.05	310.60	550	170.83	0.231	210.21	385.38
23	29.14	339.33	550	186.63	0.229	229.37	420.51
24	30.08	365.39	550	200.97	0.228	246.73	452.35
25	30.88	388.31	550	213.57	0.227	261.99	480.32
26	31.53	407.66	550	224.21	0.226	274.86	503.91
27	32.04	423.07	550	232.69	0.225	285.11	522.69
28	32.40	434.28	550	238.85	0.225	292.55	536.35
29	32.62	441.08	550	242.59	0.225	297.07	544.64
30	32.70	443.51	550	243.93	0.224	298.69	547.59

We note that biomass estimation cannot be provided for years 1 to 7, because the expected mean diameter in these years is far too low compared to the range of diameters for which the allometric equation selected is valid. Using per hectare carbon stocks, we calculate the total carbon stock in 296 ha, which is our project area. For this, we must use the data of year-wise area to be planted. We propose that the area of 296 ha will be taken up for reforestation in a phased manner over a period of three years. The areas of lands to be reforested in different years are shown in Table A13.5.

Table A13.5 Areas of lands to be reforested in different years

Year	Area (ha)
1	50
2	100
3	146
Total	296

By combining the data from Tables A13.4 and A13.5, we arrive at ex-ante estimation of carbon stocks in tree biomass within the project boundary, as summarized in Table A13.6.

Table A13.6 Ex-ante estimation of carbon stock in tree biomass within the project boundary

Project year	Area 1 (ha)	Carbon stock 1 (tCO_2e/ha)	Area 2 (ha)	Carbon stock 2 (tCO_2e/ha)	Area 3 (ha)	Carbon stock 3 (tCO_2/ha)	Carbon stock project (tCO_2e)
1	50	—	—	—	—	—	—
2	50	100	—	—	—	—	—
3	50	100	—	—	146	—	—
4	50	100	—	—	146	—	—
5	50	100	—	—	146	—	—
6	50	100	—	—	146	—	—
7	50	100	—	—	146	—	—
8	50	6.93	100	—	146	—	346.67
9	50	11.38	100	6.93	146	—	1,262.17
10	50	17.61	100	11.38	146	6.93	3,030.47
11	50	27.21	100	17.61	146	11.38	4,782.38
12	50	40.48	100	27.21	146	17.61	7,315.58
13	50	58.34	100	40.48	146	27.21	10,936.66
14	50	81.85	100	58.34	146	40.48	15,835.62
15	50	112.18	100	81.85	146	58.34	22,311.17
16	50	148.00	100	112.18	146	81.85	30,567.82
17	50	186.45	100	148.00	146	112.18	40,501.31
18	50	226.57	100	186.45	146	148.00	51,581.50
19	50	267.41	100	226.57	146	186.45	63,248.74
20	50	308.06	100	267.41	146	226.57	75,222.49
21	50	347.65	100	308.06	146	267.41	87,230.12
22	50	385.38	100	347.65	146	308.06	99,011.26
23	50	420.51	100	385.38	146	347.65	110,321.30
24	50	452.35	100	420.51	146	385.38	120,934.11
25	50	480.32	100	452.35	146	420.51	130,644.43
26	50	503.91	100	480.32	146	452.35	139,269.68
27	50	522.69	100	503.91	146	480.32	146,651.52
28	50	536.35	100	522.69	146	503.91	152,657.10
29	50	544.64	100	536.35	146	522.69	157,180.04
30	50	547.59	100	544.64	146	536.35	160,150.14

SHRUB BIOMASS IN PROJECT

Since we do not expect a significant shrub biomass in our project area once the forest has been restored to its natural state, we make the conservative choice of accounting shrub biomass as zero.

DEAD WOOD AND LITTER IN PROJECT

For the sake of simplicity, we do not wish to make field measurements to estimate dead wood and litter in our project. Instead, we use the default factors that estimate dead wood and litter biomass in the project scenario as a fraction of the above-ground tree biomass. According to the tool, dead wood and litter can be estimated as 2% and 4%, respectively, of the above-ground tree biomass, which is reflected in columns (d) and (e) of Table A13.7.

CHANGES IN SOIL ORGANIC CARBON IN PROJECT

Soil organic carbon (SOC) is likely to increase as a result of regeneration of forests because the input of organic matter to the soil will be greater under the project scenario compared to the baseline scenario. However, it is not possible to quantify the increase in SOC using the "tool for estimation of change in SOC stocks due to the implementation of afforestation and reforestation CDM project activities", because the tool is applicable only when the land-use change occurs from cropland or grassland to forested land. The baseline land use in our project is neither cropland nor grassland. Therefore, we make the conservative choice of accounting change in SOC pool as zero.

PROJECT EMISSIONS

Next, we note that the methodology only requires us to account for project emissions resulting from field burning of biomass for the purpose of site preparation. In our project, fire will not be used for site preparation, and, therefore, project emissions of our project are zero.

ACTUAL NET GHG REMOVALS

Taking into account the ex-ante estimation of changes in carbon stocks in various pools, as well as emissions resulting from implementation of the project, we estimate the actual net GHG removals as summarized in Table A13.7.

Table A13.7 Ex-ante estimation of actual net GHG removals (in tCO2e)

Project year	Carbon stock in pools/components of pools (tCO_2e)					Change in carbon stock in carbon pools	Project emissions	Actual net GHG removals
	Tree biomass	*Shrub biomass*	*Dead wood*	*Litter*	*Total*			
(a)	*(b)*	*(c)*	*(d)*	*(e)*	*(f)*	*(g)*	*(h)*	*(i)*
1	—	—	—	—	—	—	—	—
2	—	—	—	—	—	—	—	—
3	—	—	—	—	—	—	—	—
4	—	—	—	—	—	—	—	—
5	—	—	—	—	—	—	—	—
6	—	—	—	—	—	—	—	—
7	—	—	—	—	—	—	—	—
8	346.67	—	6.93	13.87	367.47	367.47	—	367.47
9	1,262.17	—	25.24	50.49	1,337.90	970.44	—	970.44
10	3,030.47	—	60.61	121.22	3,212.30	1,874.40	—	1,874.40
11	4,782.38	—	95.65	191.30	5,069.32	1,857.02	—	1,857.02
12	7,315.58	—	146.31	292.62	7,754.51	2,685.19	—	2,685.19
13	10,936.66	—	218.73	437.47	11,592.86	3,838.35	—	3,838.35
14	15,835.62	—	316.71	633.42	16,785.75	5,192.89	—	5,192.89
15	22,311.17	—	446.22	892.45	23,649.84	6,864.08	—	6,864.08
16	30,567.82	—	611.36	1,222.71	32,401.89	8,752.05	—	8,752.05
17	40,501.31	—	810.03	1,620.05	42,931.39	10,529.50	—	10,529.50
18	51,581.50	—	1,031.63	2,063.26	54,676.39	11,745.01	—	11,745.01
19	63,248.74	—	1,264.97	2,529.95	67,043.67	12,367.27	—	12,367.27
20	75,222.49	—	1,504.45	3,008.90	79,735.84	12,692.17	—	12,692.17
21	87,230.12	—	1,744.60	3,489.20	92,463.92	12,728.09	—	12,728.09
22	99,011.26	—	1,980.23	3,960.45	104,951.94	12,488.02	—	12,488.02
23	110,321.30	—	2,206.43	4,412.85	116,940.58	11,988.64	—	11,988.64
24	120,934.11	—	2,418.68	4,837.36	128,190.16	11,249.58	—	11,249.58
25	130,644.43	—	2,612.89	5,225.78	138,483.10	10,292.94	—	10,292.94
26	139,269.68	—	2,785.39	5,570.79	147,625.86	9,142.76	—	9,142.76
27	146,651.52	—	2,933.03	5,866.06	155,450.61	7,824.75	—	7,824.75
28	152,657.10	—	3,053.14	6,106.28	161,816.53	6,365.92	—	6,365.92
29	157,180.04	—	3,143.60	6,287.20	166,610.85	4,794.32	—	4,794.32
30	160,150.14	—	3,203.00	6,406.01	169,759.14	3,148.30	—	3,148.30

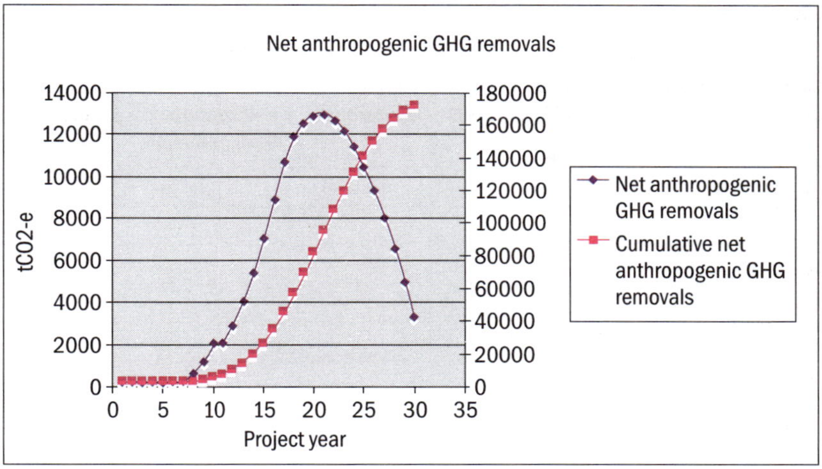

FIGURE A13.1 Net anthropogenic GHG removal over the proect period

LEAKAGE EMISSIONS

We note that the only service provided by the vacant forest lands in the baseline scenario is seasonal grazing of livestock for a few months after the monsoon season. Grazing activity will be displaced to areas within the project (by rotation, for the first two years) and then temporarily outside the project area until the regenerated trees are big enough to be safe against damage from livestock. Since the emissions from grazing activity will not increase because of displacement of grazing, the leakage emissions in our project are accounted as zero.[15]

NET ANTHROPOGENIC GHG REMOVALS

By subtracting leakage emissions and baseline removals (both of which happen to be zero in our project) from the actual net GHG removals, we obtain year-wise and cumulative net anthropogenic GHG removals of our project, as summarized in Table A13.8 and shown in the Figure A13.1. Ex-ante estimation of cumulative net anthropogenic GHG removals is also the ex-ante estimation of tCERs.

[15] However, if forested areas were to be cleared to accommodate grazing of livestock displaced from the project area, leakage emissions equal to the carbon stocks held in various carbon pools of the forested areas cleared would have been accounted.

Table A13.8 Ex-ante estimation of net anthropogenic GHG removals by sinks (tCO_2)

Project year	Baseline net GHG removals	Actual net GHG removals	Leakage emissions	Net anthropogenic GHG removals	Cumulative net anthropogenic GHG removals
1	—	—	—	—	—
2	—	—	—	—	—
3	—	—	—	—	—
4	—	—	—	—	—
5	—	—	—	—	—
6	—	—	—	—	—
7	—	—	—	—	—
8	—	367.47	—	367.47	367.47
9	—	970.44	—	970.44	1,337.90
10	—	1,874.40	—	1,874.40	3,212.30
11	—	1,857.02	—	1,857.02	5,069.32
12	—	2,685.19	—	2,685.19	7,754.51
13	—	3,838.35	—	3,838.35	11,592.86
14	—	5,192.89	—	5,192.89	16,785.75
15	—	6,864.08	—	6,864.08	23,649.84
16	—	8,752.05	—	8,752.05	32,401.89
17	—	10,529.50	—	10,529.50	42,931.39
18	—	11,745.01	—	11,745.01	54,676.39
19	—	12,367.27	—	12,367.27	67,043.67
20	—	12,692.17	—	12,692.17	79,735.84
21	—	12,728.09	—	12,728.09	92,463.92
22	—	12,488.02	—	12,488.02	104,951.94
23	—	11,988.64	—	11,988.64	116,940.58
24	—	11,249.58	—	11,249.58	128,190.16
25	—	10,292.94	—	10,292.94	138,483.10
26	—	9,142.76	—	9,142.76	147,625.86
27	—	7,824.75	—	7,824.75	155,450.61
28	—	6,365.92	—	6,365.92	161,816.53
29	—	4,794.32	—	4,794.32	166,610.85
30	—	3,148.30	—	3,148.30	169,759.14

MONITORING PLAN

In the monitoring plan, we are required to provide information on how we will carry out ex-post (actual) estimation of carbon stocks and carbon-stock changes in various carbon pools as well as project emissions and leakage emissions. We note that the methodology does not require us to monitor carbon-stock changes in the baseline. The pre-project carbon stock in tree biomass has already been estimated by us and documented in our PDD. We will use this estimate in our first monitoring report, which will be the outcome of implementation of our monitoring plan. We also note that we do not have to monitor project emissions and leakage emissions since these are accounted as zero.

We further note that changes in SOC are not required to be monitored in our project because these changes are conservatively accounted as zero. The carbon stocks and changes in carbon stocks occurring in the carbon pools of (1) below-ground biomass, (2) dead wood, and (3) litter are not required to be monitored because carbon stocks in these pools are accounted as a fixed percentage of carbon stocks in above-ground tree biomass.

Therefore, in our project the only carbon pool to be monitored is the above-ground biomass, and within this pool, we will monitor only the above-ground tree biomass since we have made the conservative choice of not accounting for the biomass of shrubs in our project.

MONITORING OF ABOVE-GROUND TREE BIOMASS

For project monitoring purposes, we will use tree allometric equations that convert tree diameter into tree biomass. We have the choice of either developing new equations or modifying existing equations that fit the data obtained from sample trees harvested from the project area. We choose the option of developing new allometric equations at the time of monitoring, so as to get an unbiased and accurate estimate of tree biomass.[16]

Since we are regenerating a mixed forest to its natural state, we will develop a generalized allometric equation by regressing tree diameter

[16] We could have made the other choice of modifying an existing equation to fit field data from sample trees harvested from within the project area. While applying an existing equation, we are allowed to modify it in such a way that it gives a conservative estimate of tree biomass but as close to the unbiased estimate as it can go.

(DBH) against field data of tree biomass across the tree species that will be present in the regenerated forest.[17]

SAMPLE PLOTS

We will apply our generalized allometric equation to all trees falling within circular sample plots having a radius of 12.61 m (that is, we will employ fixed area plots of size 0.05 ha). Sample plots within a stratum will be laid out in a systematic manner with a random start. Since no clear-cutting will take place in our project area, we choose the option of installing permanent sample plots. This means we will record the locations of plot centres and preserve these locations in database. Plot centres, however, will not be visible in field. When we are required to produce a monitoring report after the first monitoring report, we can opt for re-measuring only a fraction of plots to estimate the increase in biomass since the first monitoring report. This will ensure that our plot measurement costs are reduced.

STRATIFICATION

At the time of monitoring, we will stratify the project area into areas having (1) low, (2) medium, and (3) high above-ground tree biomass per hectare. It is not necessary to assign specific cut-off values to define where a stratum begins and where it ends. We can just make a visual inspection (or take a look at the freely available Google Earth image of our project area) and assign parcels of areas to one stratum or another. Also, a stratum does not have to consist of geographically contiguous areas. A stratum is a list (and a map) of areas that appear to have similar biomass density. Once we have firmed up the list of areas by strata and demarcated the boundaries of areas on map (based on field survey of the boundaries), these lists with maps constitute our sampling frame. To draw sample plots from a stratum, we will divide the stratum into 0.05 ha square cells. We will number these cells from 1 to *N*. To

[17] While developing species-specific allometric equations requires harvesting fewer sample trees because of a closer correlation between the DBH and tree biomass, if there are more than four or five species planted in mixed stands (rather than in species-wise blocks), it is more convenient to develop and apply a generalized equation that is valid across species. Developing a generalized equation will require us to have a larger sample (for example, 50 to 60 trees), but when we add up all the trees to be felled for each species (in the case of a number of species-specific equations), the generalized allometric equation will prove to be more economical in terms of effort. Apart from this, with a generalized equation, our plot-census fieldwork will get simplified because we will not have to identify and record tree species while recording tree diameters.

pick up n samples at random, we will generate a random cell number between 0 and N/n. We will make this cell as the starting point and then pick up every n^{th} cell from here until we reach N. The centres of the sampled cells will be the centres of our sample plots. We will pick up the latitude-longitude (or local x-y coordinates) of these plot centres from the map (these can also be calculated for us by the computer), and using a GPS, we will navigate to the plot centres in the field. At each plot centre in the field, we will drive a peg into the ground, tie one end of a tape of 12.61 m length to the peg, and swing the free end of the tape into a circle, thus defining the boundary of our sample plot. We will repeat the procedure for each stratum.

We do not propose any ex-ante stratification map in the PDD since we do not know, in advance, in what pattern the spatial distribution of biomass of the natural forest will emerge.

SAMPLE SIZE

In view of the variability of natural forest that we expect to regenerate, we assume that a sampling fraction of 1.5% will be adequate to meet the precision requirement stipulated by the methodology. Thus, in a stratum having an area of 100 ha, the number of sample plots will be 30 (so that the area sampled will be 1.5 ha or 1.5%). However, we cannot predict the variability of biomass content per hectare in the actual forest that will be emerging in our project, and hence we keep the option of determining actual sample size at the time of monitoring.

CONCLUSION

This completes the essential aspects of our afforestation and reforestation CDM project. It remains to complete the PDD form by inserting the details we have worked out above in their appropriate places and to submit our PDD to a DOE for validation and onward submission to the UNFCCC Secretariat.

Source: UNFCCC (2013)

REFERENCES AND BIBLIOGRAPHY

REFERENCES

Aggarwal, A. 2010. *Reducing Emissions from Deforestation and Degradation: issues and prospects.* New Delhi: The Energy and Resources Institute (TERI)

Asare, R. A. 2010. *Implications of the Legal and Policy Framework for Tree and Forest Carbon in Ghana: REDD opportunities scoping exercise.* Forest Trends: Washington, DC

Baalmann, P. and B. Schlamadinger. 2008. *Scaling Up AFOLU Mitigation Activities in Non-Annex I Countries. 12 June 2008—Working Paper. A report by Climate Strategies and GHG Offset Services for the Eliasch Review.* London: Climate Strategies. Details available at <http://digital.library.unt.edu/ark:/67531/metadc13692/m2/1/high_res_d/Climate_Strategies_AFOLU_mitgation.pdf>

Bayon, R., A. Hawn, and K. Hamilton. 2007. *Voluntary Carbon Markets: an international business guide to what they are and how they work. Environmental Market Insights.* London: EarthScan

Bettinger, P., M. Clutter, J. Siry, M. Kane, and J. Pait. 2009. Broad implications of Southern United States pine clonal forestry on planning and management of forests. *International Forestry Review* 11(3): 331–345

Cacho, O. 2006. Abatement and transactions costs of carbon-sink projects involving smallholders. Paper presented at the *Workshop on Climate Mitigation Measures in the Agro-forestry Sector and Biodiversity Futures,* 16–17 October 2006, Trieste, Italy, organized jointly by the Ecological and Environmental Economics (EEE) Programme, the Abdus Salam International Centre for Theoretical Physics (ICTP), UNESCO Man and the Biosphere Programme (MAB), and the International Institute for Applied Systems Analysis (IIASA)

Cacho, O. and W. Russell. 2006. Transaction and abatement costs of agroforestry projects for carbon sequestration. Details available at

<http://www.worldagroforestry.org/sea/Products/training/GroupTra/SAFODS/data/Theme5-TandA.pdf>, last accessed on 14 June 2010

Capoor, K. and P. Ambrosi. 2006. *State and Trends of the Carbon Market 2006*. Washington, DC: World Bank

Capoor, K. and P. Ambrosi. 2007. *State and Trends of the Carbon Market 2007*. Washington, DC: The World Bank. Details available at <http://web.worldbank.org/WBSITE/EXTERNAL/NEWS/0,,contentMDK:21319781~pagePK:64257043~piPK:437376~theSitePK:4607,00.html>

CBD (Convention on Biological Diversity). 2007. Biodiversity and climate change. Details available at <www.cbd.int/doc/bioday/2007/ibd-2007-booklet-01-en.pdf>

CCBA (Climate, Community, and Biodiversity Alliance). 2008. *Climate, Community and Biodiversity: project design standards*. 2nd edn. Virginia: CCBA

Chapagain, N. and M. Banjade. 2009. Community forestry and local development: experiences of the Koshi hills from Nepal. *Forest and Livelihood* 8(2): 78–92

Chenost, C., Y. Gardette, D. Julien, G. Nicolas, P. Martin, and W. Matthieu. 2010. Bringing forest carbon projects to the market. Details available at <iklim.cob.gov.tr/iklim/Files/eKutuphane/ForestryCarbonProjects.pdf>

Christian, A. 1999. Who owes who? Climate change, debt, equity and survival. Details available at <www.christianaid.org.uk//9909whoo/whoo1.htm>

CIFOR (Centre for International Forestry Research). 2001. *Capturing the Value of Forest Carbon for Local Livelihoods*. Bogor: CIFOR

CIFOR (Centre for International Forestry Research). 2005a. A/R Clean Development Mechanism project activities: legal framework in Indonesia. Details available at <http://repository.ipb.ac.id/bitstream/handle/123456789/35359/Carbon%20Brief%2006_CDM%20Legal%20Aspect.pdf?sequence=1>

CIFOR (Centre for International Forestry Research). 2005b. *A/R Clean Development Mechanism Project Activities: project cycle*. Bogor: CIFOR

CIFOR (Centre for International Forestry Research). 2009. *USAID-CIFOR-ICRAF Project: assessing the implications of climate change for USAID Forestry Programme*. Bogor: CIFOR

CMR (Carbon Mechanisms Review). 2013. Can forestry projects play a role in the carbon market? *Carbon Mechanisms Review,* Issue 2, July-September 2013. Details available at <www.jiko-bmu.de/files/basisinformationen/application/pdf/cmr-2013-august-englisch-web.pdf>

CORE (Carbon Offset Research and Education). 2010. Mandatory & Voluntary Offset Markets. Details available at <www.co2offsetresearch.org/policy/MandatoryVsVoluntary.html>

Cornland, D. 2006. *CDM Business Issues: describing a project for buyers.* Sweden: Cornland International Environment and Development Initiatives

Davis, P. 2000. *Carbon Forestry Projects in Developing Countries: legal issues and tools.* Washington, DC: Forest Trends Details available at <www.regjeringen.no/en/dep/fin/campaign/Carbon-Neutral-Norway.html?id=479281>

Doucher, D. 2008. *Out of the Woods: a realistic role for tropical forests in curbing global warming.* Cambridge: UCS Publications

EcoSecurities. 2010. The forest carbon offsetting report 2010. Details available at <www.ecosecurities.com/Registered/ForestCarbonOffsettingReport2010.pdf>

FAO (Food and Agriculture Organization). 2007. *State of the World's Forest.* Rome: FAO

FAO (Food and Agriculture Organization). 2010a. *Carbon Finance Possibilities for Agriculture, Forestry and Other Land Use Projects in a Smallholder Context.* Rome: FAO

FAO (Food and Agriculture Organization). 2010b. *Forests and Climate Change in the Asia–Pacific Region. Forests and Climate Change Working Paper 7.* Rome: FAO

FSI (Forest Survey of India). 1996. *Volume Equations for Forests of India, Nepal, and Bhutan.* New Delhi: FSI, Ministry of Environment and Forests

FSI (Forest Survey of India). 1999. *State of Forest Report.* Dehradun: Ministry of Environment and Forests

Fujiwara, N. and E. Christian. 2010. *Understanding India's Climate Agenda. CEPS Policy Brief No. 206.* Brussels, Belgium: Centre for European Policy Studies

Gairola, S., N. M. Shariff, A. Bhatt, and C.P. Kala. 2010. Influence of climate change on production of secondary chemicals in high altitude medicinal plants: issues needs immediate attention. *Journal of Medicinal Plants Research* 4(18): 1825–1829

GoI (Government of India). 2006. *National Environmental Policy*. New Delhi: GoI. Details available at <http://envfor.nic.in/sites/default/files/introduction-nep2006e.pdf>

Graves, J. and D. Reavey. 1996. *Global Environmental Change: plants, animals and communities*. London: Longman Group Ltd

Grubb, M. 2003. The economics of the Kyoto Protocol. *World Economics* 4(3): 143–189

Grubb, M. L., T. Brewer, S. Misato, H. Robert, and F. Dora. 2009. *Climate Policy and Industrial Competitiveness: ten insights from Europe on the EU emissions trading system*. Washington, DC: The German Marshall Fund

Guariguata, M. R., J. Cornelius, B. Locatelli, C. Forner, and G. Sánchez-Azofeifa. 2010. *Mitigation Needs Adaptation: tropical forestry and climate change*. Bogor: Centre for International Forestry Research

Gustavsson, L., K. Pingoud, and R. Sathre. 2006. Carbon dioxide balance of wood substitution: comparing concrete and wood-framed buildings. *Mitigation and Adaptation Strategies for Global Change* 11(3): 667–691

Hamilton, K., B. Ricardo, T. Guy, and H. Douglas. 2007. *State of the Voluntary Markets 2007: picking up the steam*. Washington, DC: New Carbon Finance and Ecosystem Marketplace

Hamilton, K., M. Sjardin, M. Peters-Stanley, and T. Marcello. 2010. Building bridges: state of the voluntary carbon markets 2010. Details available at <http://moderncms.ecosystemmarketplace.com/repository/moderncms_documents/vcarbon_2010.2.pdf>

Hamilton, K., R. Bayon, G. Turner, and D. Higgins. 2007. *State of the Voluntary Carbon 2006: picking up steam*. Washington, DC: The Ecosystem Marketplace. Details available at: <www.ecosystemmarketplace.com/documents/acrobat/StateoftheVolunatryCarbonMarket18July_Final.PDF>

Hamilton, K., S. Ryan, S. Emma, and W. Sissel. 2008. Offsetting emissions: a business brief on the voluntary carbon market. Washington, DC: Ecosystem Marketplace and Business for Social Responsibility

Harris, N., S. Brown, S.C. Hagen, A. Baccini, and R. Houghton. 2013. *Progress Toward a Consensus on Carbon Emission from Tropical Deforestation*. Washington, DC: Warnock International and Words Hole Research Centre

Hawn, A. 2005. Horses for courses: volunatry vs. CDM carbon projects in Mexico. Details available at <www.ecosystemmarketplace.com/pages/dynamic/article.page.php?page_id=2509&section=home&eod=1>

IPCC (Intergovernmental Panel on Climate Change). 2001. *Climate Change 2001: synthesis report. A Contribution of Working Groups I, II, and III to the Third Assessment Report of the Integovernmental Panel on Climate Change*, edited by R. T. Watson and the Core Writing Team. Cambridge, UK and New York, USA: Cambridge University Press. 398 pp

IPCC (Intergovernmental Panel on Climate Change). 2001a. *Climate Change 2001: the scientific basis. Contribution of Working Group I to the Third Assessment Report of the Intergovernmental Panel on Climate Change*, edited by J. T. Houghton, Y. Ding, D. J. Griggs, M. Noguer, P. J. van der Linden, X. Dai, K. Maskell, and C. A. Johnson. Cambridge, UK and New York, USA: Cambridge University Press. 881 pp

IPCC (Intergovernmental Panel on Climate Change). 2001b. *Climate Change 2001: impacts, adaptation, and vulnerability. Contribution of Working Group II to the Third Assessment Report of the Intergovernmental Panel on Climate Change*, edited by J. J. McCarthy, O. F. Canziani, N. A. Leary, D. J. Dokken, and K. S. White. Cambridge, UK and New York, USA: Cambridge University Press

IPCC (Intergovernmental Panel on Climate Change). 2003. *Good Practice Guidance for Land-use, Land-use Change, and Forestry*. Kanagawa, Japan: Institute for Global Environmental Strategies

IPCC (Intergovernmental Panel on Climate Change). 2007. *IPCC Fourth Assessment Report. Climate Change 2007 (AR4)*. Cambridge, UK and New York, USA: Cambridge University Press

Jagger, P., E. O. Sills, K. Lawlor, and W. D. Sunderlin. 2010. *A Guide to Learning about Livelihood Impacts of REDD+ Projects*. Bogor, Indonesia: Centre for International Forestry Research

Kala, C. P. 2010a. Status of an indigenous agro-forestry system in changing climate: a case study of the middle Himalayan region of Tehri Garhwal, India. *Journal of Forest Science* 56(8): 373–380

Kala, C. P. 2010b. *Medicinal Plants of Uttarakhand: diversity, livelihood and conservation*. New Delhi: Biotech Books. 188 pp

Kala, C. P. 2010c. Ethnobotanical and ecological approaches for conservation of medicinal and aromatic plants. *Acta Horticulturae* 860: 19–26

Kanninen, M., D. Murdiyarso, F. Seymour, A. Angelsen, S. Wunder, and L. German 2007. *Do trees grow on money? The implications of deforestation research for policies to promote REDD*. Bogor, Indonesia: Centre for International Forestry Research. 61 pp

Kauppi, P. and R. Sedjo. 2001. Technological and economic potential of options to enhance, maintain, and manage biological carbon reservoirs and geo-engineering. In *Climate Change 2001: mitigation*. Cambridge, UK and New York, USA: Cambridge University Press

Kelman, I. and J. W. Jennifer. 2009. Climate change and Small island Developing States: a critical review. *Ecological and Environmental Anthropology* 57: 245–256

Korppoo, A., J. Linda U. Johannes, V. Antto, and L. Alex. 2009. *Towards a new global climate regime?* Helsinki: The Finnish Institute of International Affairs

Krolik, T. 2006. The Argentine carbon fund helps developers dance the dance. Ecosystem Marketplace. Details available at <www.ecosystemmarketplace.com/pages/dynamic/article.page.php?page_id=4371&section=home&eod=1>

Kumar, R., J. Parag, and T. Shalini. 2008. Economic impact of climate change on Mumbai, India. *Regional Health Forum* 12(1): 38–42

Lamb, D., P. D., Erskine, and J. A. Parrotta. 2005. Restoration of degraded tropical forest landscapes. *Science* 310: 1628–1632

Leemans, R. and B. Eickhout. 2004. Another reason for concern: regional and global impacts on ecosystems for different levels of climate change. *Global Environmental Change* 14: 219–228

Levina, E. and D. Tirpak. 2006. *Adaptation to Climate Change: key terms*. Paris: Organization for Economic Cooperation and Development. Details available at <www.oecd.org/env/cc/36736773.pdf>

Lichtenfeld, M. 2007. Improving the supply of carbon sequestration services in Panama. *Journal of Sustainable Forestry* 25(1/2): 43–73

Locatelli, B. and L. Pedroni. 2003. Rewarding carbon sequestration in forest plantation projects: How the CDM could be attractive to small stakeholders? Paper presented at the *2nd Henry A. Wallace Inter-American Scientific Conference,* 19–21 March 2003, Turrialba, Costa Rica

Lomborg, B. 2001. *The Skeptical Environmentalist.* Cambridge: Cambridge University Press

Manguiat, Z., S. Maria, R. Verheyen, M. Jens, and S. Gerald. 2005. *Legal Aspects in the Implementation of CDM Forestry Projects.* Switzerland and Cambridge: International Union for Conservation of Nature (IUCN) IUCN Environmental Law Centre

Mansourian, S., A. Belokurov, and P. J. Stephenson. 2009. The role of forest protected areas in adaptation to climate change. *Unasylva* 60(231/232): 63–69

Markit Environmental Registry. 2009. Beginner's guide to the voluntary carbon market. Markit Group Limited. Details available at <www.katoombagroup.org/documents/cds/uganda_2011>

Mawdsley, J. R., R. O'Malley, and D. S. Ojima. 2009. A review of climate-change adaptation strategies for wildlife management and biodiversity conservation. *Conservation Biology* 23(5): 1080–1089

Merger, E. 2010. Status and future of afforestation and reforestation (A/R) carbon sector. Details available at <www.carbonpositive.net/viewFile.as>

Minang, P. A., B. Swallow, and V. Meadu (eds). 2008. *REDD strategies for high-carbon rural development. ASB Policy Brief No. 11.* Nairobi, Kenya: ASB Partnership for the Tropical Forest Margins

MoEF (Ministry of Environment and Forests). 2004. *India's Initial Communication to the United Nations Framework Convention on Climate Change.* New Delhi: MoEF

MoEF (Ministry of Environment and Forests). 2010. *REDD+: India's forests and REDD+.* New Delhi: MoEF.

Nabuurs, G. J., O. Masera, K. Andrasko, P. Benitez-Ponce, R. Boer, M. Dutschke, E. Elsiddig, J. Ford-Robertson, P. Frumhoff, T. Karjalainen, O. Krankina, W. A. Kurz, M. Matsumoto, W. Oyhantcabal, N. H. Ravindranath, M. J. Sanz Sanchez, and X. Zhang. 2007. Forestry. In *Climate Change 2007: mitigation,* edited by B. Metz, O. R. Davidson, P. R. Bosch, R. Dave, and L. A. Meyer. Cambridge, UK and New York, USA: Cambridge University Press [Contribution of Working

Group III to the Fourth Assessment Report of the Intergovernmental Panel on Climate Change]

Neeff, T. and S. Henders. 2007. Guidebook to markets and commercialization of forestry CDM projects. The Tropical Agricultural Research and Higher Education Centre (CATIE). Details available at <www.katoombagroup.org/~katoomba/documents/tools/GuidebooktoMarketsandCommercializationofCDMforestryProjects.pdf>

Neeff, T., E. Lisa, D. Imme, and F. Jan. 2007. *Update on Markets for Forestry Offsets*. Washington, DC: The Tropical Agricultural Research and Higher Education Center (CATIE)

Nigel, M., T. Maria, R. Vargas, and J. Smith 2002. Can forest-protection carbon projects improve rural livelihoods? Analysis of the Noel Kempff Mercado Climate Action Project, Bolivia. *Mitigation and Adaptation Strategies for Global Change* 7: 323–337

Pagiola, S. and B. Bosquet. 2009. Estimating the costs of REDD at the country level. World Bank. Details available at <www.forestcarbonpartnership.org/sites/forestcarbonpartnership.org/files/Documents/PDF/REDD-Costs-22.pdf>

Parker, C., J. Brown, J. Pickering, E. Roynestad, N. Mardas, and A. W. Mitchell. 2009. *The Little Climate Finance Book: a guide to financing options for forests and climate change*. John Krebs Field, Oxford: Global Canopy Programme. Details available at <www.globalcanopy.org/sites/default/files/lcfb_en%20(1).pdf>

Parmesan, C. and G. Yohe. 2003. A globally coherent fingerprint of climate change impacts across natural systems. *Nature* 421: 37–42

Pearson, T., W. Sarah, and B. Sandra. 2006. *Guidebook for the Formulation of Afforestation and Reforestation Projects under the Clean Development Mechanism*. Yokohama, Japan: International Tropical Timber Organization

Philibert, C., E. Jane, and P. Jacek. 2007. *Carbon Capture and Storage in the CDM*. Paris: Organization for Economic Cooperation and Development. Details available at <www.iea.org/publications/freepublications/publication/CCS_in_CDM.pdf>

Rainforest Alliance. 2010. *My Earth*. New York: Rainforest Alliance

Ravindranath, N. H. 2006. Climate change, forests and India: CO_2 emissions, impacts, mitigation, and global negotiations. In *Climate change and India*, edited by anonymous. 174 pp.

Ravindranath, N. H. 2007. Mitigation and adaptation synergy in forest sector. *Mitigation and Adaptation Strategies for Global Change* 12: 843–853

Ravindranath, N. H., N. V. Joshi, R. Sukumar, and A. Saxena 2006. Impact of climate change on forests in India. *Current Science* 90(3): 354–361

Ravindranath, N. H., N. V. Joshi, R. Sukumar, I. K. Murthy, and H. S. Suresh. 2003. Vulnerability and adaptation to climate change in the forest sector. In *Climate Change and India: vulnerability assessment and adaptation*, edited by P. R. Shukla, S. K. Sharma, N. H. Ravindranath, A. Garg, and S. Bhattacharya, pp. 227–262. Hyderabad: Universities Press

Rawles, S. 2008. *REDD Myths: a critical review of proposed mechanisms to reduce emissions from deforestation and degradation in developing countries*. Netherlands: Friends of Earth International, 43 pp

Richards, M. and S. N. Panfil. 2010. *Manual for Social Impact Assessment of Land-based Carbon Projects, Version 1*. Washington, DC: Forest Trends, Climate, Community, and Biodiversity Alliance, Rainforest Alliance and Fauna, and Flora International

Robertson, T. 1998. *Guide to the Kyoto Protocol*. Amsterdam, The Netherlands: Greenpeace International

Robledo, C. and C. Forner. 2005. *Adaptation of Forest Ecosystems and the Forest Sector to Climate Change. Forests and Climate Change Working Paper 2*. Rome: Food and Agriculture Organization, Swiss Agency for Development and Cooperation, and Swiss Foundation for Development and International Cooperation, Inter-cooperation

Sala, O. E., F. S. Chapin III, J. J. Armesto, E. Berlow, J. Bloomfield, R. Dirzo, E. Huber-Sanwald, L. F. R. B. Huenneke Jackson, A. Kinzig, R. Leemans, D. M. Lodge, H. A. Mooney, M. Oesterheld, N. LeRoy Poff, M. T. Sykes, B. H. Walker, M. Walker, and D. H. Wall. 2000. Global biodiversity scenarios for the year 2100. *Science*, 287: 1770–1774

Sathaye, J., W. Makundi, C. L. P Dale, and K. Andrasko. 2007. GHG mitigation potential, costs and benefits in global forest: a dynamic partial equilibrium approach. *Energy Journal* 3(special issue): 127–172

Sathaye, J., P. Shukla, and N. H. Ravindranath. 2006. Climate change, sustainable development and India: global and national concerns. *Current Science* 90(3): 314–324

Satyanarayana, M. 2004. How forest producers and rural farmers can benefit from the Clean Development Mechanism. In *Proceedings of the Workshop on Forests for Poverty Reduction: opportunities with Clean Development Mechanism, environmental services and biodiversity,* edited by H. C. Sim, S. Appanah, and Y. C. Youn. Bangkok, Thailand: Food and Agriculture Organization. [27–29 August 2003, Seoul, Korea, organized jointly by Seoul National University, Asia Pacific Association of Forestry Research Institutions, Forestry Research Support Programme for Asia and the Pacific, Korea Forest Economics Society, Northeast Asia Forest Forum]

Sawyer, S., M. Meinshausen, and S. Guilbeault. 2003. *Sinks in the CDM: after the climate, biodiversity go down the drain? An analysis of the CDM sinks agreement at CoP-9.* Milan, Italy: Greenpeace International

Scholes, R. and S. Linder. 1998. Forests. In *Handbook on Methods for Climate Change Impact Assessment and Adaptation Strategies,* edited by J. Feenstra, I. Burton, J. Smith, and R. Tol. Ámsterdam: UNEP-Vrije Universiteit

Searle, N. 2010. Avoiding the courts. *Carbon Finance* 18 February 2010

Seymour, F. 2011. Maintaining forests for adaptation and mitigation. Details available at <www.worldresourcesreport.org/responses/no-regrets-maintaining-forests-adaptation-and-mitigation>

Smit, B. and W. Skinner. 2002. Adaptation options in agriculture to climate change: a topology. *Mitigation and Adaptation Strategies for Global Change* 7: 85–114

Smith, J. and S. J. Scherr. 2002. *Forest Carbon and Local Livelihoods: assessment of opportunities and policy recommendations.* Indonesia: Center for International Forestry Research

Stern, N. 2007. *Stern Review on the Economics of Climate Change.* Cambridge: Cambridge University Press

Susan, F. R. 2001. Global climate change: the Kyoto Protocol. Congressional Research Service. Details available at <www.au.af.mil/au/awc/awcgate/crs/rl30692.pdf>

Swallow, B., M. van Noordwijk, S. Dewi, D. Murdiyarso, D. White., J. Gockowski, G. Hyman, S. Budidarsono, V. Robiglio, V. Meadu, A. Ekadinata, F. Agus, K. Hairiah, P. Mbile, D. J. Sonwa, and S. Weise. 2007. *Opportunities for Avoided Deforestation with Sustainable Benefits. An Interim Report by the ASB Partnership for the Tropical*

Forest Margins. Nairobi, Kenya: ASB Partnership for the Tropical Forest Margins

Tewari, D. D. and J. Campbell. 1996. Increased development of non-wood products in India: some issues and concerns. *Unasylva* 47: 26–31

UNEP (United Nations Environment Programme) and Ecosecurities. 2007. Guidebook to financing CDM projects. Details available at <www.cd4cdm.org/publications/FinanceCDMprojectsGuidebook.pdf>

UNFCCC (United Nations Framework Convention on Climate Change). 2003. Methodological Issues. Land use, land use change and forestry: definitions and modalities for including afforestation and reforestation activities under Article 12 of the Kyoto Protocol for the first commitment period. Options paper on modalities for addressing socio-economic and environmental impacts, including impacts on biodiversity and natural ecosystems. Eighteenth Session of the Subsidiary Body for Scientific and Technological Advice, Bonn, 4–13 June 2003. Details available at <http://.int/resource/docs/2003/sbsta/07.pdf>

UNFCCC (United Nations Framework Convention on Climate Change). 2010. CDM glossary. Details available at <http://cdm..int/Reference/Guidclarif/glos_CDM.pdf>

UNFCCC (United Nations Framework Convention on Climate Change). 2010. Report of the Subsidiary Body for Scientific and Technological Advice on its thirty-first session, held in Copenhagen from 8 to 12 December 2009. Details available at <www.preventionweb.net/files/12841_081.pdf>

UNFCCC (United Nations Framework Convention on Climate Change). 2013. Afforestation and Reforestation Projects under the Clean Development Mechanism: a reference manual. Bonn, Germany: UNFCCC

Walker, S. M., T. R. H. Pearson, P. Munishi, and S. Petrova. 2008. Carbon market opportunities for the forestry sector of Africa. Paper presented at the *16th Session of the African Forestry and Wildlife Commission, Khartoum, Sudan, 17–24 February 2008*. Khartoum, Sudan: Winrock International. Details available at <www.winrock.org/ecosystems/files/Winrock_FAO_Carbon_opportunities_in_Africa.pdf>

Walter, G. R., E. Post, P. Convey, A. Menzel, C. Parmesan, T. J. C. Beebee, J. M. Fromentin, O. Hoegh-Guldberg, and F. Bairlein. 2002. Ecological responses to recent climate change. *Nature* 416: 389–395

Wara, M. 2007. Is the global carbon market working? *Nature* 445: 595–596

World Bank. 2008. Scaling up carbon finance in India. Background paper. India: strategies for low carbon growth. Details available at <http://envfor.nic.in/sites/default/files/LCGCarbonJune2008.pdf>

Zwick, S. 2010. 2010: the year in forest carbon. Ecosystem Marketplace. Details available at <www.ecosystemmarketplace.com/pages/dynamic/article.page.php?page_id=7920&section=news_articles&eod=1>

BIBLIOGRAPHY

Bettinger, P. 2006. Measures of change in projected landscape conditions. *Journal of Forest Planting* 12(2): 39–47

Carbon Positive. 2011. Green light for California cap and trade. Details available at <www.carbonpositive.net/viewarticle.aspx?articleID=2220>

CCX (Chicago Climate Exchange). CCX offset Report. Details available at <http://theccx.com/docs/offsets/Reports/CCX_Offsets_Report_Vol1No5_Sept_Dec.pdf>

CIFOR (Centre for International Forestry Research). 2009. *Simply REDD: CIFOR's guide to forests, climate change and REDD*. Bogor, Indonesia: CIFOR

Hindustan Times. 2011. India's own emission reducing mechanism. *Hindustan Times*, New Delhi, 27 January 2011. Details available at <www.hindustantimes.com/India-s-own-emission-reducing-mechanism/H1-Article1-655381.aspx>

IPCC (Intergovernmental Panel on Climate Change). 2007. *Climate Change 2007: impacts, adaptation and vulnerability. Contribution of Working Group II to the Fourth Assessment Report of the Intergovernmental Panel on Climate Change*, edited by M. L. Parry, O. F. Canziani, J. P. Palutikof, P. J. van der Linden, and C. E. Hanson. Cambridge, UK and New York, USA: Cambridge University Press. 976 pp

IPCC (Intergovernmental Panel on Climate Change). 2007. *Climate Change 2007: the physical science basis. Contribution of Working Group I to the Fourth Assessment Report of the Intergovernmental Panel on*

Climate Change, edited by S. Solomon, D. Qin, M. Manning, Z. Chen, M. Marquis, K. B. Averyt, M. Tignor, and H. L. Miller. Cambridge, UK and New York, USA: Cambridge University Press. 996 pp

Mery, G., R. Alfaro, M. Kanninen, and M. Lobovikov (eds). 2005. *Forests in the Global Balance: changing paradigms. IUFRO World Series Vol. 17.* Vienna, Austria: International Union of Forest Research Organizations. 318 pp

MoEF (Ministry of Environment and Forests). 1992. *National Conservation Strategy and Policy Statement on Environment and Development.* New Delhi: MoEF. Details available at <http://envfor.nic.in/sites/default/files/introduction-csps.pdf>

Noriko, F. 2010. Sectoral approaches to climate change: what can industry contribute? Details available at <www.ceps.eu/book/sectoral-approaches-climate-change-what-can-industry-contribute>

Palm, C., S. A. Vosti, P. A. Sanchez, and P. J. Ericksen. 2005. Slash and burn agriculture: the search for alternatives. New York: Columbia University

Point Carbon. 2007. Canada seen small buyer of CERs through 2010. Details available at <www.pointcarbon.com/article21969-70.html?articleID=21969&categoryID=470>

Revenga, C. 2000. Will there be enough water. World Resources Institute. Details available at <http://earthtrends.wri.org/pdf_library/feature/wat_fea_scarcity.pdf>

Richards, M. and S. Panfil. 2010. *Manual for Social Impact Assessment of Land-Based Carbon Projects (Version 1.0): Part I—Core Guidance for Project Proponents.* Washington, DC: Forests Trends

Sari, A. P. and S. Meyers. 1999. *Clean Development Mechanism: perspectives from developing countries.* US Environmental Protection Agency

SCBD (Secretariat of the Convention on Biological Diversity). 2009. *Connecting Biodiversity and Climate Change Mitigation and Adaptation: report of the Second Ad Hoc Technical Expert Group on Biodiversity and Climate Change. Technical Series No. 41.* Montreal: SCBD

Taiyab, N. 2006. *Exploring the Market for Voluntary Carbon Offsets.* London: International Institute for Environment and Development

Takacs, D. 2009. Forest carbon—law and property rights. Arlington, VA, USA: Conservation International

UNDESA (UN Department of Economic and Social Affairs). 2009. Indicators of sustainable development. Details available at <www. un.org/esa/sustdev/natlinfo/indicators/methodology_sheets/poverty/ without_electricity.pdf>

UNFCCC (United Nations Framework Convention on Climate Change). 2001. Climate change information sheets. Details available at <http://. int/resource/docs/publications/infokit_2002_en.pdf>

UNFCCC (United Nations Framework Convention on Climate Change). 2008. Bali Action Plan. *Report of the Conference of the Parties on its thirteenth session,* held in Bali from 3 to 15 December 2007. Details available at <http://.int/resource/docs/2007/cop13/eng/06a01.pdf>

United Republic of Tanzania. 2007. *A Handbook for Clean Development Mechanism (CDM) Project Activities in Tanzania*. Tanzania: Division of Environment

Verner, D. 2010. *Reducing Poverty, Protecting Livelihoods and Building Assets in a Changing Climate: social implications of climate change in Latin America and the Caribbean*. Washington, DC: World Bank

World Bank. 2004. *Sustaining Forests: a development strategy*. Washington, DC: World Bank

WRI (World Resources Institute). 2000. *A Guide to World Resources 2000–2001. People and Ecosystems: the fraying web of life*. Washington, DC: WRI

GLOSSARY

Actual net greenhouse gas removals by sinks: The sum of the verifiable changes in carbon stocks in the carbon pools within a project boundary that are attributable to an afforestation and reforestation or small-scale afforestation and reforestation Clean Development Mechanism (CDM) project activity or programme of activities (PoA) (afforestation and reforestation), as applicable, minus any increase in anthropogenic greenhouse gas (GHG) emissions by sources (measured in carbon dioxide equivalents) within the project boundary that is caused by the implementation of the afforestation and reforestation or small-scale afforestation and reforestation CDM project activity or PoA (afforestation and reforestation), as applicable.

Adaptation: Adjustments in natural or human systems that try to reduce the damage caused by climate change or to exploit the benefits.

Additional: A small-scale afforestation or reforestation project activity under the CDM is additional if the actual net GHG removals by sinks are increased above the sum of the changes in carbon stocks in the carbon pools within the project boundary that would have occurred in the absence of the registered small-scale afforestation or reforestation project activity under the CDM.

Additionality tool: Tool developed by the CDM Executive Board (and adapted by the afforestation and reforestation working group for afforestation and reforestation projects) for the purpose of demonstrating additionality.

Afforestation and reforestation CDM modalities and procedures: Modalities and procedures for afforestation and reforestation project activities under the CDM, as adopted by the CMP (Conference of the Parties Serving as the Meeting of the Parties to the Kyoto Protocol) in decision 5/CMP.1 and including any subsequent amendments.

Afforestation and reforestation CDM project activity: An afforestation or reforestation measure, operation, or action that aims to achieve net anthropogenic GHG removals by sinks, whether as a whole project or as a part of a project.

Afforestation: Direct human-induced conversion of land that has not been forested for a period of at least 50 years to forested land through planting, seeding, and/or the human-induced promotion of natural seed sources. Afforestation is also defined as planting new forest on land that has not previously supported forest.

Annex I Party: A party that is included in Annex I to the Convention or a Party that has made a notification under Article 4, paragraph 2(g) of the Convention.

Anthropogenic emissions: GHGs associated with human activity such as deforestation or forest degradation from logging.

Baseline: Emissions or removals of GHGs that would occur without the introduction of the project.

Biofuels: Fuels from renewable biological sources in either solid or liquid form. Biofuel crops associated with deforestation include palm oil, sugar cane, and soya.

Biomass: Non-fossilized and biodegradable organic materials originating from plants, animals, and microorganisms. Biomass also includes products, by-products, residues, and waste from agriculture, forestry, and related industries; non-fossilized and biodegradable organic fractions of industrial and municipal wastes; and gases and liquids recovered from the decomposition of non-fossilized and biodegradable organic material.

Cap and trade: A two-part regulatory system in which the "cap" is a government-imposed limit on carbon emissions, and the "trade" is a government-created market to buy and sell GHG credits. Companies that generate less than the allowed emissions can sell credits, allowing others to emit more gases than the cap allows.

Carbon dioxide: A gas occurring naturally in the atmosphere, which is produced as a by-product of combustion of fossil fuels and biomass, with changes in land use, and during industrial processes.

Carbon markets: Financing bodies and mechanisms that can exchange carbon credits generated from verified Reducing Emissions from Deforestation and forest Degradation (REDD) activities. These are of two types: "voluntary markets" (formed under bilateral agreement between the trading parties) and "compliance market" (legally regulated to meet emission reduction target under multilateral agreements).

Carbon pool: A system that can accumulate or release carbon. Examples include forest biomass, wood products, soils, and atmosphere; also defined as organic materials containing carbon.

Carbon sequestration: Uptake and storage of carbon. Trees absorb carbon dioxide and release oxygen during photosynthesis, and also store carbon in their biomass.

Carbon sink: The process or mechanism through which carbon dioxide is absorbed and stored in organic matters such as forests, oceans, and soil.

Carbon stock: Quantity of carbon in a given pool(s) per unit area.

Carbon tax: A surcharge levied on consumers who emit carbon dioxide into the atmosphere.

Carbon trading: Transaction of verified or certified carbon credits generated from REDD in monetary terms.

CDM modalities and procedures: Modalities and procedures for a CDM as adopted by the CMP in the annex to decision 3/CMP.1 and including any subsequent amendments.

Certification: The written assurance by a Designated Operational Entity (DOE) that an afforestation and reforestation CDM project activity or PoA achieved the net anthropogenic GHG removals by sinks since the start of the project, as verified.

Certified emission reductions: A unit issued pursuant to Article 12 and requirements there under, as well as the relevant provisions in the CDM modalities and procedures. A certified emission reduction (CER) is equal to 1 tonne of carbon dioxide equivalent, calculated using global warming potentials defined by decision 2/CP.3 or as subsequently revised in accordance with Article 5 of the Kyoto Protocol.

Clean Development Mechanism: A scheme that helps industrialized countries meet their Kyoto Protocol emission targets, in addition to reducing their own emissions, in two ways: they can invest in reducing GHG emissions or in enhancing GHG sinks in developing countries. In Article 12 of the Kyoto Protocol, CDM is defined as: "The purpose of the clean development mechanism shall be to assist Parties not included in Annex I in achieving sustainable development and in contributing to the ultimate objective of the Convention, and to assist Parties included in Annex I in achieving compliance with their quantified emission limitation and reduction commitments under Article 3".

Climate change: A change in the mean meteorological parameters that define climate or their variability, including temperature, rainfall, and wind speed.

Co-benefits: Benefits from implementing REDD schemes, in addition to reducing GHG emissions, such as poverty alleviation, biodiversity protection, and improvement in forest governance.

Component project activity: A component project activity (CPA) is a single measure, or a set of interrelated measures under a PoA, to reduce GHG emissions by sources or result in net anthropogenic GHG removals by sinks, applied within a designated area defined in the baseline methodology(ies).

Conference of the Parties: A decision-making body comprising the parties that have ratified the UNFCCC.

Convention: The United Nations Framework Convention on Climate Change (UNFCCC).

Coordinating/managing entity: An entity authorized by all participating host country designated national authorities (DNAs) involved in a particular PoA and nominated in the modalities of communication statement as the entity that communicates with the board and the secretariat, including on matters relating to the distribution of CERs, temporary CERs (tCERs), or long-term CERs (lCERs), as applicable.

CoP/MoP: Conference of the Parties (CoP) to the UNFCCC and Meeting of the Parties (MoP) to the Kyoto Protocol.

CPA design document: The document prepared by the coordinating/managing entity (CME) that sets out in detail, in accordance with the CDM rules and requirements, the CPA to be undertaken. The form of CPA design document (CPA-DD) and guidelines on preparing the CPA-DD are publicly available on the UNFCCC CDM website.

Crediting period for afforestation and reforestation CDM project activities: The period during which net anthropogenic GHG removals by sinks are verified and certified by a DOE for issuing lCERs or tCERs. The crediting period shall start with the start of the afforestation and reforestation CDM project activity and continue during its operational lifetime. The board, at its 21st meeting, clarified that the provisions mentioned in paragraphs 12 and 13 of decision

17/CP.7 do not apply to afforestation and reforestation CDM project activities. An afforestation and reforestation CDM project activity starting after 1 January 2000 can also be validated and registered after 31 December 2005 as long as the first verification of the project activity occurs after its date of registration. If the start of the crediting period coincides with that of the project activity, the projects starting 2000 onwards can accrue tCERs/lCERs as of the starting date. Project participants may choose between two options for the length of a crediting period: (1) fixed crediting period or (2) renewable crediting period, as defined in paragraph 23(a) and (b) of the afforestation and reforestation CDM modalities and procedures.

Deforestation: Change of forested land to treeless land.

Designated National Authority: UNFCCC National Focal Point takes this role.

Designated Operational Entity: An independent, approved, verifying organization of the Kyoto Protocol; until fully accredited, it is known as applicant entity (AE).

Ecosystem services: Benefits provided to humans by an ecosystem. For example, forests provide food, water, timber, and fibre; regulate climate, floods, disease, and water quality; and also provide recreational, aesthetic, and spiritual benefits.

Ecosystem: A community of organisms and their physical environment.

Eligibility (land): To be eligible for an afforestation and reforestation CDM project, land must have been deforested before 31 December 1989 and continuously without forest since that date. Project participants shall follow the latest procedures to define the eligibility of lands, as available at: http://cdm.unfccc.int/Reference/Procedures (UNFCCC 2010).

Executive Board: The governing body of the CDM.

Forest (afforestation and reforestation; small-scale afforestation and reforestation): A "forest" covers a minimum of 0.05–1.0 ha of land, with more than 10%–30% tree crown cover (or equivalent stocking level); the trees should grow to reach a minimum height of 2–5 m at maturity in situ. A forest may consist either of closed forest formations, where trees of various storeys and undergrowth cover a high proportion of the ground, or of open forest. Forest includes young natural stands and all plantations that have not yet reached a crown density of 10%–30% or a height of 2–5 m, as well as part of the forest area that is temporarily unstocked by harvesting or natural causes but is expected to revert to forest. A party not included in Annex I may host an afforestation and reforestation CDM project activity if it has selected and reported the parameters chosen for the definition of "forest" to be used in the afforestation and reforestation project activities under the CDM to the Executive Board through its DNA for the CDM.

GHG sink: Any process or mechanism to remove a GHG, an aerosol, or a precursor of a GHG from the atmosphere.

Global warming: The increase in the Earth's average temperature year to year, which leads to climate change.

Greenhouse effect: A process in which gases such as carbon dioxide prevent the energy that is generated by the sun and is radiated back from the surface of the earth in the form of heat to escape the earth's atmosphere.

Greenhouse gases: Gases in the atmosphere (both natural and anthropogenic), mainly water vapour, carbon dioxide, nitrous oxide, methane, and ozone, that absorb and emit radiation, causing the greenhouse effect.

Host party: A party not included in Annex I to the Convention on whose territory the CDM project activity is physically located. A project activity can have several host parties if it is located in more than one country. A host party shall meet the requirements for participation, as defined in paragraphs 28–30 of the CDM modalities and procedures, to register for the project activity.

Intergovernmental Panel on Climate Change: A joint UN Environment Programme–World Meteorological Organization body responsible for providing the scientific and technical foundations for the UNFCCC.

Issuance of CERs: Instruction by the Executive Board to the CDM Registry Administrator to issue a specified quantity of CERs for a project activity into the pending account of the Executive Board in the CDM registry, in accordance with paragraph 66 and Appendix D of the CDM modalities and procedures.

Issuance of tCERs or lCERs (afforestation and reforestation): Instruction by the Executive Board to the CDM Registry Administrator to issue a specified quantity of lCERs or tCERs for an afforestation and reforestation CDM project activity into the pending account of the Executive Board in the CDM registry, in accordance with CDM modalities and procedures and sections J and K and Appendix D of the afforestation and reforestation CDM modalities and procedures. Upon the issuance of tCERs or lCERs, the CDM Registry Administrator shall, in accordance with CDM modalities and procedures, promptly forward the tCERs or lCERs to the holding accounts of project participants involved, in accordance with their request, having deducted the quantity of tCERs or lCERs corresponding to the share of proceeds to cover administrative expenses for the Executive Board.

Kyoto Protocol: An international agreement covering the period 2008–12 to slow climate change. Under the Kyoto Protocol, industrialized countries agreed to reduce their collective GHG emissions by 5.2% below the 1990 levels.

Leakage: The loss of carbon outside the boundaries of the project as a result of project activities or the increase in GHG emissions by sources outside the boundary of an afforestation and reforestation CDM project activity. It is measurable and attributable to the afforestation and reforestation CDM project activity. It can also be defined as the incident that occurs when a reduction of emissions in one area leads to an increase in emissions in another area. For example, a REDD project protects forest in one area, but involves in deforestation activities elsewhere. Leakage is also known as emission displacement.

Liability: The obligation of the REDD implementing project or country to ensure that the credited emission reductions are permanent.

Long-term CER: A unit issued pursuant to Article 12 of the Kyoto Protocol for an afforestation and reforestation CDM project activity, which expires at the end of the crediting period of the project activity for which it was issued. It is equivalent to 1 tonne of carbon dioxide.

Mitigation: Actions to reduce GHG emissions and enhance carbon sinks to curb climate change.

Monitoring methodology: The method used by project participants to collect and archive all relevant data necessary for the implementation of the monitoring plan.

Monitoring plan: The plan that sets out the methodology to be used by project participants or CMEs for the monitoring of and for the verification (by DOEs) of the amount of reductions of anthropogenic emissions by sources or removals by sinks of GHGs achieved by the CDM project activity or PoA, as applicable.

Monitoring report: A report prepared by a project participant that sets out the GHG emission reductions or net GHG removals of an implemented registered CDM project activity or PoA for a particular monitoring period.

Monitoring: The collection and archiving of all relevant data necessary for estimating or measuring the net anthropogenic GHG removals by sinks during the crediting period. For more information on the monitoring plan, please refer to paragraph 25 of the afforestation and reforestation CDM modalities and procedures.

Nationally Appropriate Mitigation Action: Voluntary or mandatory action by a developing country to reduce its carbon emissions in line with its economic, environmental, social, and political contexts.

Net anthropogenic GHG removals by sinks: The actual net GHG removals by sinks minus the baseline net GHG removals by sinks minus leakage.

Non-Annex I Parties: Parties to the Convention that are not included in Annex I to the Convention.

Party involved: A party that provides a written approval.

Payments for environmental services: Schemes where beneficiaries of ecosystem services pay those who ensure that the services continue.

Peat: An accumulation of partially decayed vegetation matter formed in wetlands, including bogs, moors, and peat swamp forests.

Permanence: A measure of the anticipated longevity of carbon sequestered as part of a CDM activity. It is also the duration and reversibility of reduced GHG emissions.

Planted forest: Wooded land where trees have been established through planting or seeding.

PoA design document: The document prepared by the CME of a PoA, which sets out in detail, in accordance with the CDM rules and requirements, the PoA to be undertaken. The form of PoA design document (DD) and guidelines on preparing the PoA-DD are publicly available on the UNFCCC CDM website.

Primary forest: Wooded land of native species largely untouched by human activities, where ecological processes are not disrupted.

Programme of activities: A voluntary coordinated action by a private or public entity that coordinates and implements any policy/measure or stated goal (that is, incentive schemes and voluntary programmes). The goal involves anthropogenic GHG emission reduction or net anthropogenic GHG removal through sinks that are additional to any that would occur in the absence of the PoA, via an unlimited number of CPAs.

Project activity (afforestation and reforestation CDM): An afforestation and reforestation measure, operation, or action to achieve net anthropogenic GHG removal by sinks. The term "project activity" is used in the Kyoto Protocol and the CDM modalities and procedures, as opposed to "project". An afforestation and reforestation CDM project activity could be identical with a component or an aspect of a project undertaken or planned.

Project boundary for afforestation and reforestation project activities: The "project boundary" geographically delineates the afforestation and reforestation CDM project activity under the control of the project participants. An afforestation and reforestation CDM project activity may involve more than one discrete area of land.

Project design document: The document prepared by the project participant of a CDM project activity, which sets out in detail, in accordance with the CDM rules and requirements, the CDM project activity to be undertaken. The form of the project design document (PDD) and guidelines on preparing the PDD are publicly available on the UNFCCC CDM website.

Project idea note: A document prepared early in project planning that defines the type of project, the anticipated location, cost, and potential sequestration.

Project participants: According to CDM modalities and procedures, a project participant is (1) a party that has indicated to be a project participant or (2) a private and/or public entity authorized by a party, involved in a CDM project activity. As per Appendix D of the CDM modalities and procedures, project participants have the sole right to take decision on the distribution of CERs from a CDM project activity. Project participants need to write to the Executive Board, through the secretariat, following the "modalities of communication", as indicated at the time of registration or as subsequently altered. If a project participant wishes to stay away from decision making on the distribution of CERs, it should communicate the same to the Executive Board, through the secretariat, at the latest when the request regarding the distribution is made.

REDD+: Broader REDD frameworks that include forest conservation, sustainable forest management, or enhancement of forest carbon stocks to encourage greater participation in REDD and to reward countries that are already protecting their forests.

Reduced impact logging: Planned and carefully controlled tree felling to minimize its impact on the surrounding environment and also to reduce the carbon emissions generated by logging activities.

Reforestation: Establishment of forest plantations in areas regarded as former forest lands; in other words, re-establishing forested land, which became non-forested land, through planting, seeding, and/or human-induced promotion of natural seed sources. For the first commitment period, reforestation activities will be limited to those lands that did not contain forest on 31 December 1989.

Registration: The formal acceptance of a validated project activity as a CDM project activity by the Executive Board. Registration is the prerequisite for the verification, certification, and issuance of CERs related to that project activity.

Sequestration: The process of increasing the carbon stock in an ecosystem.

Small-scale afforestation and reforestation CDM modalities and procedures: Modalities and procedures for small-scale afforestation and reforestation CDM project activities under the CDM, as adopted by the CMP in decision 6/CMP.1, including any subsequent amendments.

Small-scale afforestation and reforestation CDM project activity: An afforestation and reforestation measure, operation, or action to attain net anthropogenic GHG removals of less than 8 kilotonnes of carbon dioxide by sinks per year; however, this definition is applicable when the average projected net anthropogenic GHG removals by sinks do not exceed 8 kilotonnes of carbon dioxide per year for each verification period and are developed or implemented by low-income communities and individuals, as determined by the host party. If in a small-scale afforestation and reforestation project activity under the CDM, the net anthropogenic GHG removals by sinks exceed 8 kilotonnes of carbon dioxide equivalent per year, the excess removals will not be eligible for the issuance of tCERs or lCERs. The Kyoto Protocol and the CDM modalities and procedures use the term "project activity" as opposed to "project". A small-scale afforestation and reforestation CDM project activity could, therefore, be identical with a component or aspect of a project undertaken or planned.

Stakeholders: The public, including individuals, groups, or communities, affected, or likely to be affected, by the proposed CDM project activity or actions, leading to the implementation of such an activity.

Starting date of an afforestation and reforestation CDM project activity (afforestation and reforestation): An afforestation and reforestation CDM project activity starting after 1 January 2000 can also be validated and registered after 31 December 2005 as long as the first verification of the project activity occurs after its date of registration. If the start of the crediting period coincides with that of the project activity, the projects starting 2000 onwards can accrue tCERs/ lCERs as of the starting date. This clarification was provided by the board in paragraph 64, of its 21st meeting report and stipulates that provisions of paragraphs 12 and 13 of decision 17/CP.7 do not apply to afforestation and reforestation CDM project activities.

Stern review: Sir Nicholas Stern's report in 2006 for the British government that examines the effect of climate change on the world economy; although this review was not the first of its kind, it is perhaps the most influential.

Temperate: Mean annual temperature between 0 and 20°C.

Temporary CERs: A unit issued pursuant to Article 12 of the Kyoto Protocol for an afforestation and reforestation CDM project activity under the CDM, which expires at the end of the commitment period following the one during which it was issued. It is equivalent to 1 tonne of carbon dioxide. If project participants have chosen to issue tCERs to address non-permanence, they have to make a request to the Executive Board for the issuance of tCERs equal to the verified amount of net anthropogenic GHG removals by sinks achieved by the afforestation and reforestation CDM project activity under the CDM since its start.

Tropical: Mean annual temperature greater than 20°C.

United Nations Framework Convention on Climate Change: the 1992 treaty calling for the stabilization of GHG concentrations in the atmosphere at a level that would not "dangerously" affect the earth's climate.

Validation: The process of independent evaluation of a proposed small-scale/large-scale afforestation and reforestation CDM project activity under the CDM by a DOE against the requirements of small-scale/large-scale afforestation and reforestation project activities under the CDM as set out in decision 3/CMP.1, 4/CMP.1, 5/CMP.1, and 6/CMP.1, its annex and relevant decisions of the CoP/MoP, on the basis of the PDD. The DOE may perform validation, verification, and certification for a small-scale/large-scale afforestation and reforestation project activity under the CDM or for bundled small-scale/large-scale afforestation and reforestation project activities under the CDM.

Verification: The periodic independent review and ex-post determination by the DOE of the net anthropogenic GHG removals by sinks achieved, since the start of the project, by an afforestation and reforestation CDM project activity under the CDM. Certification is the written assurance by a DOE that an afforestation and reforestation CDM project activity under the CDM achieved the net anthropogenic GHG removals by sinks since the start of the project, as verified.

INDEX

Note: *f* denotes figure and *t* denotes table

About the Authors

Dr H S Gupta is Director of Tribal Empowerment and Livelihood Project, Jharkhand. Prior to this, he served as Senior Professor of Technical Forestry at the Indian Institute of Forest Management (IIFM), Bhopal, with research focus on collaborative working in forestry projects, plantation forestry, management of wildlife, and climate change in forestry. He is an Indian Forest Service officer of the Jharkhand cadre (1986 batch). Dr Gupta has done his PhD from BIT Mesra, MTech from IIT Delhi, and BS from Allahabad University in Agricultural Engineering. He has worked extensively in different districts of Bihar and Jharkhand, carrying out various models of plantation/afforestation, forest and wildlife management, and has analysed the different dimensions of their efficacy and efficiency. His current interest and research experience are in climate change related issues to forestry, tribal and rural development, conflict resolution, policy, and laws in natural resource management, sustainability of natural resources and certification aspects of forestry.

His other areas of interest include JFM, protected area management, watershed management, non-timber forest product management, tribal development, rural development, application of remote sensing and GIS in forest and wildlife management, conservation laws and policies, climate change and CDM in forestry, afforestation and plantation management, conflict resolution in NRM, and governance and policy issues in forestry. He can be reached at hsgupta@iifm.ac.in.

Dr M. Yadav is Associate Professor of Marketing Management at IIFM, Bhopal. He holds a PhD in international marketing and an MBA in marketing. He has industry experience of more than nine years and academic experience of 15 years. At IIFM, he teaches on corporate environmental responsibility and green marketing, supply chain management, and international business and forest certification. Besides teaching, Dr Yadav conducts specialized training programs for senior officers of the Indian Forest Service, industry executives, and researchers on green marketing strategies and forest certification. He has published a number of research papers in refereed journals and in national and international conferences.

Dr Yadav is currently conducting research related to peoples' brand, green business strategies, non-timber forest product marketing, information system and enterprise development, eco-labelling, and forest certification. He can be reached at manmohan@iifm.ac.in.

Dr M Verma is a Professor of Environment and Developmental Economics at IIFM, Bhopal. She has 27 years of work experience with post-doctoral research work at the University of California (Berkeley) and the University of Massachusetts (Amherst), USA. She has a PhD in industrial economics and is also a Lead International Fellow and Fulbright Environmental Leadership awardee. She has more than 40 publications in international and national journals/books to her credit and has conducted projects on valuation, green accounting, and payment for ecosystem services with the support of the World Bank, United Nations Environment Programme, European Union, Department for International Development, Asian Development Bank, International Institute for Environment and Development, Winrock International, and various ministries of Government of India—Ministry of Environment and Forests, Ministry of Statistics and Programme Implementation, Ministry of Health, Ministry of Defence, National Forestry Commission, and the 13th Finance Commission of India. She has also contributed to the world-acclaimed projects on millennium ecosystem assessment and the economics of ecosystems and biodiversity.

Her areas of expertise and research interests include economic valuation of ecosystem services, green accounting, livelihoods economics, conservation and environmental finance—economics of climate change, carbon finance and developing markets for ecosystem services, ecosystem modelling of wetlands, forests, and water resources, and policy analysis. She can be reached at mverma@iifm.ac.in.

Dr A David is an Assistant Professor of Sociology and Community Development at IIFM, Bhopal. He has a PhD in sociology from the Indian Institute of Technology Bombay and an MA in social work with specialization in urban and rural community development from the Tata Institute of Social Sciences Mumbai. His areas of interest include social impact assessment, protected area management and management effectiveness evaluation, human–wildlife interface conflicts, and ecotourism. He can be reached at adavid@iifm.ac.in.

Dr U K Sarma is an Assistant Professor of Sociology and Community Development at IIFM, Bhopal. He has a PhD and MSc in Anthropology from University of Delhi. At IIFM, he teaches courses on natural resource management for rural development; people, park, and wildlife; society and polity; NGO management; and civil society and corporate environmental responsibility. Prior to IIFM, he worked in the Wildlife Trust of India and WWF India and was part of more than 20 wildlife conservation action projects across 12 states of India, including two autonomous district councils in north-east India supported by the World Society for Protection of Animals, International Fund for Animal Welfare, US Fish and Wildlife Services, and Japan Tiger and Elephant Fund. He is a member of Indian Anthropological Association, North-East India Council for Social Science Research, and IUCN Task Force on Cultural and Spiritual Values of Protected Areas.

His areas of interest include cultural ecology, environmentalism, livelihoods and wildlife conservation, human–wildlife conflict, indigenous knowledge and biodiversity conservation, participatory conservation (with special emphasis on north-east India), gender dimension in environmental management, culture of poverty, and social change in Indian Society. He can be reached at ujjal@iifm.ac.in.

Dr C P Kala is an Assistant Professor of Ecosystem and Environment Management at IIFM, Bhopal. He has a PhD in Forestry from the Forest Research Institute, Dehradun and an MSc in Botany from the HNB Garhwal University, Srinagar. Some of his earlier associations were with the Wildlife Institute of India, Dehradun; GB Pant Institute of Himalayan Environment and Development, Almora; and National Medicinal Plants Board, Ministry of Health and Family Welfare, Government of India, New Delhi. Dr Kala has surveyed various biogeographical areas extensively for natural resource management, biodiversity conservation, and ecosystem and environment management. He has published 5 books and over 140 research papers and articles in various national and international journals and magazines. He is in the advisory and editorial board of a dozen of reputed national and international journals, which include the *International Journal of Forestry Research*, USA; *American Journal of Plant Sciences*, USA; *International Journal of Ecology*, USA; *Journal of Biodiversity*, India; *Studies on Ethno-medicine*, India; and *Medicinal and Aromatic Plants*, USA. Besides, he is the founder Executive Editor of the *International Journal of Medicinal Plants*. He is a recipient of the prestigious "ICFRE Award for Excellence" for his outstanding contribution in the field of Forest Conservation (Biodiversity and Ecology).